Intelligent Manufacturing

The EIU Series

This innovative series of books is the result of a publishing collaboration between Addison-Wesley and the Economist Intelligence Unit. Our authors draw on the results of original research by the EIU's skilled research and editorial staff to provide a range of topical, information-rich and incisive business titles. They are specifically tailored to the needs of international executives and business education worldwide.

Titles in the Series

Daniels — *Information Technology: The Management Challenge*

Irons — *Managing Service Companies: Strategies for Success*

Egan and McKiernan — *Inside Fortress Europe: Strategies for the Single Market*

Manser — *Control from Brussels*

Mazur and Hogg — *The Marketing Challenge*

Paliwoda — *Investment Opportunities in Eastern Europe*

Intelligent Manufacturing

Lynn Underwood

ADDISON-WESLEY PUBLISHING COMPANY

Wokingham, England • Reading, Massachusetts • Menlo Park, California • New York
Don Mills, Ontario • Amsterdam • Bonn • Sydney • Singapore
Tokyo • Madrid • San Juan • Milan • Paris • Mexico City • Seoul • Taipei

This book uses material drawn from the report, *The New Manufacturing: Minimal IT for Maximum Profit,* published jointly by the Economist Intelligence Unit and Reed Business Publishing.

Cover designed by Pencil Box Ltd, Marlow, Buckinghamshire
incorporating photograph by Kerry Lawrence
and printed by The Riverside Printing Co. (Reading) Ltd.
Text designed by Valerie O'Donnell.
Line diagrams drawn by Margaret Macknelly Design, Tadley.
Typeset by Meridian Phototypesetting Limited, Pangbourne.
Printed in Great Britain at the University Press, Cambridge.

First printed 1993.

ISBN 0-201-62418-4

British Library Cataloguing in Publication Data
A catalogue record for this book is available from the British Library.

Library of Congress Cataloging in Publication Data applied for.

Foreword

This book has arisen from an Economist Intelligence Unit Report which Ingersoll Engineers originally co-authored. Expanded and updated, it is now available to a much wider audience.

It is a forward-looking book for a new generation of managers, who are satisfied with nothing less than competing as equals on their own particular world stage.

Its chief subject is increasing the cash-creating potential of manufacturing business through judicious use of information technology (IT). This book presents a pragmatic approach, in clear language, to achieving effective integration of information management. This achievement has to be based on a sound understanding of all aspects of business competitiveness.

The focus of management in western manufacturing industry has shifted from 'survival' to 'performance'. The standards are set by worldwide competitors. The only acceptable performance level is 'world class'. Advances in IT, manufacturing technology and management practices are yielding results. The next big step is integration – and the medium for integration is information.

The IT industry now has the technology and skills to provide modern managements with sophisticated means to collect, interpret, analyse and present information for decision making. Managers need to be more selective in choosing such tools, rejecting those which do not have a direct impact on business performance, but investing confidently in those that do.

Today, even more than when this material was first published, this competence is vital to every effective manager and would-be manager, and is a challenge that they cannot afford to ignore.

Brian Small
Managing Director
Ingersoll Engineers Limited

Contents

Executive summary

(1) The four factors of production: labour, raw materials, capital and land, have been joined by what has become virtually the most important factor in the economics of industry – information. The use of information, in its widest sense, to develop a corporate strategy and maintain a market lead, and the use of technology to manage that information, has been the cornerstone of the success of Far Eastern manufacturers. The West has yet to learn how to use IT to its full potential.

(2) New technology and fiercer global competition has forced many companies to rethink their methods and structures. Western manufacturers have often got it wrong by trying to graft ever more complex technology on to inadequate organizations and by trying to achieve computer-integrated manufacturing without the infrastructure to support such a bold move.

(3) A company needs to analyse every aspect of its organization, and the competitive environment in which it operates, before analysing its technology needs. A company must define what it is aiming to achieve through its manufacturing strategy before putting IT to work. Simplification is the key to success. Technology, if not managed correctly, can multiply existing complexities.

(4) Companies should always be prepared to look for help when planning and implementing a manufacturing system. Sources of advice, if not free, are plentiful. Few companies can see for themselves what changing technology will mean for their organization.

(5) Lack of investment is seen as the biggest factor preventing most western manufacturers from being world class. Two factors seem to worry managers the most – how to get value for money from IT and how to control IT costs

once they've got it. The role of the IT manager/director has now changed from purely that of a technician to that of a business manager. IT strategy must be inextricably linked to corporate strategy for the investment to pay off.

(6) Risk management is an important part of today's hi-tech manufacturing environment. Physical and intellectual property has to be protected. Risks have to be assessed in the planning stage of all manufacturing decisions. Then risks have to be limited as much as possible by various administrative and technical controls. Contingency plans for disaster recovery have to be set in place.

(7) Quality, reliability and cost are all interconnected. Companies should seek to maximize value through a commitment to quality at all levels of the company structure and manufacturing operation.

(8) Change in any organization requires careful planning to maximize the benefits and minimize disruption. Companies should also be sufficiently aware to be able to anticipate change thrust upon them by external forces.

(9) To achieve the status of world class manufacturer requires the ultimate effort of aiming for excellence in every facet of the organization and also in all business relationships with suppliers, contractors, vendors and agents. There is also a need constantly to review the logistical base of the organization.

(10) The new manufacturing philosophy requires a radical change in accounting practice but, more importantly, in its whole orientation. It requires an integrated partnership in which accounting follows, rather than leads, the manufacturing and business strategy.

(11) IT-intensive manufacturing has revolutionary consequences for the traditional roles of people in production, making their value as an asset greater. All workers are turned into information workers. This new role, with its heavy emphasis on flexibility, puts a premium on intensive learning and training at all levels. Management structures need to be simplified, reducing traditional hierarchies dramatically. Managers, too, have to learn to work more flexibly. Lines of demarcation between departments and traditional craft skills have been broken down by technology.

(12) Modern IT permits an almost unlimited range of possible manufacturing architectures. In the information age, product and process are intimately connected. But technology should be subservient to, and a significant enhancer of, people and products, not the other way about. To look at technology as a substitute for people and a cheapener of products is essentially negative.

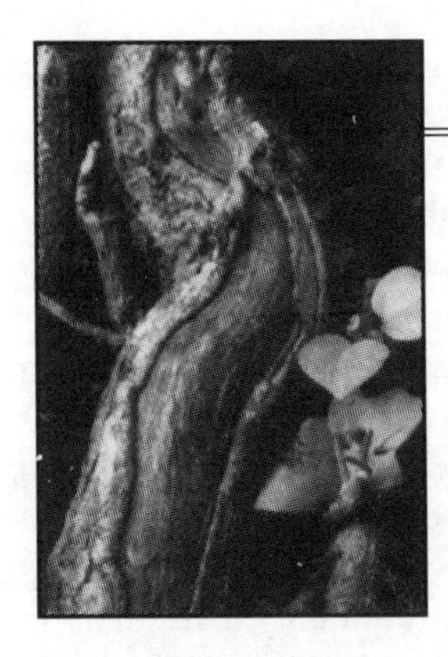

Introduction: How to use IT and survive

Western manufacturers are at a crossroads. They can either respond to the competitive challenge from the Far East or roll over and die. They do not *have* to die – as the performance of a few world class manufacturing firms in both Europe and the USA clearly demonstrates. For most, reversing the decline requires learning the Japanese lesson, going back to manufacturing basics and rethinking the entire factory process.

A central challenge is how to use the vital enabling tool of information technology (IT). Too often in the West, IT becomes a competitive handicap instead of a competitive gain. Its potential to speed up the flow of information is wasted as firms simply superimpose it on antiquated existing products and processes, automating complexity instead of simplicity, rigid instead of flexible systems. Computerized manufacture can make a good firm better; it cannot by itself turn a bad firm into a good one.

This book describes how a manufacturing firm can exploit technology and prosper. There is no mystery about the process, although for most firms it is a radical break. Chapters 1 and 2 set out the theory of 'the new manufacturing': the historical and environmental conditions which spawned it and the 'frugal' philosophy behind it. Chapter 3 emphasizes the need for firms to develop a *manufacturing strategy* before they even think of applying IT-based processes, and outlines the three key steps in the move towards world class systems: simplify–integrate–automate.

Few firms possess all the resources required to operate the transformations essential for the new manufacturing. That means an opportunity rather than a handicap. Experience or change is the only route to becoming a flexible 'intelligent' organization which welcomes continuous improvement. Chapter 4 describes the resources available to help the firm make the most of the learning process.

After acquiring an understanding of the competitive challenges, analysing its own corporate needs and developing a strategy, perhaps with help from outside sources, it is then time for a company to make the decision to invest in IT. Not an easy decision, as Chapter 5 demonstrates – the value of IT has proved consistently difficult for some companies to evaluate and therefore much of western industry still perceives it as a cost, rather than an investment.

Once they have invested in IT, few companies manage it as a resource. For that reason they fail to recognize threats to it. Yet the possible risks are serious and become more so with increased computerization. Risk management (Chapter 6) is a little-understood, powerful incentive to adopt a robust, frugal approach to IT in manufacturing.

Approaching world class manufacturing requires not only investment in technology but an investment in a philosophy which pushes a company into a level of excellence above that of its competitors. Total quality management (TQM), as discussed in Chapter 7, has proved the answer for many companies, but it requires a disciplined organizational structure which many western companies seem unable to achieve because of the radical change involved to their current procedures. But change is inevitable: if it is not imposed from inside the company, it will be imposed from outside. Changing markets, changing economies, fast-changing technology are all unavoidable. The subject of Chapter 8 is instigating, anticipating and managing change.

Having got the infrastructure right by developing the right organizational systems and philosophies, it is time to concentrate on the ultimate goal – world class manufacturing. This is examined in Chapter 9. Essentially it means aiming for excellence in every aspect of the company – not a job for the faint-hearted.

Chapter 10 examines the fact that many western firms are badly served by their accounting systems; a key information framework for tactical and strategic control. Old style accountancy gives too much weight to audit data, uses inadequate measures for daily control and fails to account for strategic purpose in investment. The secret is to use indicators which reflect business aims (quality, inventory control, overall productivity) and adapt investment criteria to quantify the intangibles.

Traditional manufacturing relies on the division of labour to standardize and speed tasks. Incentive schemes are linked to these arrangements to provide motivation. The penalty: increasing alienation of the workforce and, eventually, lagging productivity. Chapter 11 shows how the new manufacturing demands integrated jobs, both on the factory floor and throughout the management levels. Fortunately, by simple organizations and short information loops, modern manufacturing also makes such jobs desirable, with dramatic gains in shop floor motivation and creativity. Make the most of a step change in the man–machine interface by first winning over sceptical middle managers, linking all implementation with on the job training and rewarding people in a way which supports the new manufacturing.

In Chapter 12, technology is revisited. In the post-technological age, the emphasis switches from technology itself to what you want to do with it. The aim of the new manufacturing approach is to achieve more with less – lean production – substituting information for effort. Within this framework, consider IT as a resource like any other, with associated costs as well as benefits. Information strategy is the key to continuous manufacturing improvement, which then becomes the model and driver for the rest of the firm, pushing product and process enhancement through the white collar functions. In a world of overcapacity, world class manufacturing is only a step towards the ultimate goal of just-in-time (JIT) business; nil wasted time, nil wasted quantity, nil wasted quality.

1 The issues

Western manufacturing is changing fast. Good labour, productivity and 'the white heat of technology' are no longer particularly relevant. Product design and process management skills are the new hallmarks of success.

There are four factors of production in the classical economics of the firm: labour, raw materials, capital and land. Success has been achieved by the optimum combination of these variables to produce a product which is cheaper than its rivals and which maximizes the entrepreneurs' profits over time. Whether this mechanistic model was ever adequate to explain company behaviour in the real world is arguable. It certainly is not adequate now, when a fifth variable has become as important as all the others put together. This fifth factor of production is information.

Information in its broadest sense: management know-how as well as accounting, product and market intelligence, has always been an essential part of manufacturing. But for two interlinked reasons, its role has steadily grown, until it is now the single most important resource at the disposal of the manufacturing firm. The first reason is the dramatic development of information related technology (IT) itself, which feeds the appetite for information in the very act of providing the means of satisfying it. The second is the equally dramatic transformation of the global economic topography since World War II, affecting every company in the world. By the same token, the importance and complexity of information's role also increases its destructive power for those who ignore it or get it wrong, as western manufacturers are now beginning to testify.

Companies have succumbed, during their love affair with the computer, to a near-fatal confusion of information itself with the information technology tools for gathering, manipulating and storing it. As a result, computers have done as much harm as good for many manufacturing firms, encouraging obsession with technology and distracting attention from fundamental underlying needs in industrial organization, relationships and finance. 'We have so much technology available that it's choking us' a US manager told the *Wall Street Journal* (1988). A new generation of IT tools is touted as the saviour of western manufacturing, with encouragement from not only suppliers but also governments to 'automate or liquidate'. But is IT the solution to every manufacturing problem? There is an urgent need to stand back from the technology and reassess its role in the factory of the present as well as the factory of the future.

Global competition

To understand the importance of the proper use of information, we look first at the changing competitive landscape. Western assumptions of manufacturing superiority die hard. The Old World was the centre of economic power in the nineteenth century just as the New World has been for much of the twentieth. Britain was the cradle of the first industrial revolution, as the USA was of the second. The very terms in which manufacturing was discussed were western. Most of the economics, all the management theory (such as it is) and, until very recently, all the management techniques were invented and developed in Europe and the USA. The factory, the steam engine, the production line, the division of labour, the functional management structure and the microchip: the history of the first age of manufacturing is by definition that of the western, 'developed' world.

The rise of the East

All that is now changing. The competitive centre of gravity has shifted east, partly because of the dramatic spread of information-based technology and its refusal to recognize national boundaries. The rise of the Japanese economy since World War II is as well known as the British economy's decline. Even with a constantly appreciating currency, Japan has been increasing its share of world trade as fast as the UK and the USA have been losing theirs. At least until the currency adjustments of the mid-1980s, its manufacturing productivity has been increasing faster and its unit labour costs more slowly than its immediate competitors. Followed at a respectful though narrowing distance by South East Asia's 'little dragons' (South Korea, Taiwan, Hong Kong), Japan has not only successively conquered shipbuilding, motorbikes, automobiles, machine tools, construction equipment, cameras, audio, TV and video, semiconductors, robots and computers both small and large; it has turned them into global products, marketable with only minor modifications in the market 'triad' of Europe, North America and the Pacific basin. In all these global market segments, Japanese and some South Korean companies manufacture at least as well as, and in some cases manifestly better than, their western rivals. They, not the West, are making the running.

There is no sign of any let-up on the fierce competitive pressures. One powerful reason is that the soaring yen, like the successive increases in energy costs in the 1970s, has increased rather than decreased the stakes in the high-spending spiral in which the big Japanese manufacturers find themselves caught. Already harried by overcapacity at home, within months of 1985's currency accord between the Group of Five they were reported to be writing off a generation of unpaid-for existing plants and investing in even newer factories. The drive is to launch the new products to win the export volumes, even at present exchange rates, which will gain the market share and justify the raised stakes in yet more expensive plant and still shorter product life-cycles the next time round.

The end of the home market

For western firms, the consequence of the rise of an export-oriented South East Asia is that, in a relatively open trading system, there is no longer any hiding place from international competition. In the past, the conventional strategy was for a UK company to establish a firm platform in the home market from which to launch into exports, which in many cases were treated as a kind of luxury; an easy way of soaking up surplus production. With its huge home market, the US company hardly needed to bother about exports at all.

With one or two exceptions, Japan being the most notable because of tariff and other barriers, it no longer makes sense to speak of home markets. Increasingly, for a company to retain its share of a local market means that it must meet world standards of competitiveness. Import penetration figures confirm this. To sell cars in the UK, Rover must manufacture to Toyota's levels of quality and value for money. As Sir Edwin Nixon, former chairman of IBM UK, noted (IBM, 1987), purely national productivity and wage differentials, and even economies of scale, are losing their importance as yardsticks of industrial performance:

> 'The criteria for success will be international standards of efficiency. And the technology necessary for achieving that efficiency will be

available to every industrial concern across the globe at the same cost of investment.'

Adopting innovatory process technology involves a trade-off in which the company bets the high risk and monetary cost of being first against the possibility of gaining real competitive advantage. The newer the technology the greater the risk that it will fail, either technically or financially; the older and more tried the technology the less the chance of stealing a march on the opposition.

In the past, decisions to install numerically controlled (NC) machine tools in the late 1950s and 1960s and computer aided design (CAD), computer numerically controlled (CNC), automated warehouses and programmable controllers in the 1970s, hardly seemed strategic. Like the computer doing the payroll in the finance department, they aimed to optimize performance or quality in one particular business area (the drawing office, the machine shop, the stores) without reference to other departments. Now these discrete initiatives look rather different: like halting, piecemeal and sometimes even contradictory steps towards the logical conclusion of a factory run by computer.

The present state of manufacturing

Progress even in these limited areas has been uneven. British machine tool manufacturers, for instance, have long lamented UK industry's slow uptake of advanced technology. The USA is an extensive user of CAD, which it developed, but has been curiously less swift to develop it into computer aided manufacture (CAM): in the five years to 1986, only 18% of its new machine tools were CNC, the essential building blocks of computer aided manufacturing, as opposed to 55% in Japan. Japan is also the champion implementer in robotics, claiming 93,000 industrial robots in operation in 1985 against 20,000 in the USA, 8800 in West Germany and 3200 in the UK (Owen, 1987). On the other hand, the USA has the lion's share of the self-styled fully-fledged computer-integrated manufacturing (CIM) installations running, several of which are in the defence industries.

Together with depressing trends in manufacturing trade balances, world export shares and import penetration, such figures tend to evoke panic and alarm in the West. There is now quasi-unanimity at office level that advanced manufacturing technology is the only way for industrial companies not only to catch up with but to leapfrog both their sophisticated Japanese rivals and the low-wage competition in South Korea, Taiwan and elsewhere. Advanced manufacturing is the subject of official research programmes in Europe and almost every industrialized nation in the world.

Japan and the West compared

Research of this kind dwells lengthily on the potential benefits of high-tech manufacturing. But it pays less attention to the risks, and in particular to the implications, of the contrast between the Japanese and western patterns of automation investment. The difference is significant and revealing. The US, and to a lesser extent European, attempt to leap directly to CIM represented a discontinuity, an abrupt departure from previous manufacturing theory and practice. This ambitious once-for-all approach was in strong contrast to the Japanese practice of building upwards from inside the business: first process rationalization and simplification preparing the ground for the individual CNC machine tool, moving on to their combination in flexible manufacturing cells and only then (if justified) their linking in the fully computer integrated factory. This methodical progress strongly suggests that, unlike western firms, the large Japanese companies are already in possession of what might be termed an advanced manufacturing philosophy. Far from being a radical break with the past, as in the West, for Japanese manufacturers concepts such as CIM (in so far as they are important in themselves) are a logical extension of the present, making them both simpler to envisage and easier to achieve.

Fortunately, these differences provide important pointers to the implementation of the new technologies for less advanced companies. The variety of international experience, both success and failure, makes it possible for manufacturers for the first time to take a rational view of the strategic issues and their investment implications, avoiding the twin pitfalls of overconfidence in technical solutions and paralysis in the face of their complexity. The growing confidence is partly, of course, to do with gradually increasing familiarity with, and reliability of, the technological tools. And development continues: the next few years will certainly see progress at all levels of information technology in the factory, as well as new kinds of factory to exploit them. More importantly, however, there is beginning to be a better understanding of the strengths and limitations of the new tools, and of how to use them. Not before time, computers are being stripped of their witchdoctor's mystery. As NEDO (1985) put it:

> 'Now flexible automation is commercially viable, and it is possible to achieve high volume productivity with medium volume, wider variety output. The performance of companies using modern technology depends on how far and how well they implement it, not on the technology itself.'

Case study: Hewlett Packard 1988

In a radical departure, Hewlett Packard (HP) started making cheap computer terminals, 'the lowest priced high quality terminals in the world'. Its new strategy was a test case for the ability of western manufacturers to compete internationally in mass markets. HP spent two years analysing the decision. What effect would cheaper terminals have on its reputation as a 'boutique' manufacturer of expensive products in small volumes? Its customers certainly used low-cost terminals in certain circumstances to complement their expensive ones: but 'HP quality', in the words of one manager, 'could not be compromised'.

HP took apart the Taiwan-built Wyse 30, a low-cost competitor, to understand its engineering and labour costs. Conclusion: Wyse competed on labour costs alone, not clever engineering. So if HP could design terminals that were easy to build, it could be competitive. The decision to make rather than buy meant tackling labour costs issue head on, but gave advantages in control and quality. HP's three-point plan follows.

- *Design for manufacturing ability.* R&D and engineering brainstormed. All expenditures were scrutinized. When the products went into production, HP had reduced design changes by a factor of ten.
- *Involve suppliers.* Rather than buy special parts, HP asked suppliers for the least expensive part which would satisfy most needs. The onus was on the supplier to meet quality standards. HP got to know more about suppliers' costs. Further innovations: packaging suppliers stuck on the HP labels. A plastics supplier devised a snap shut case which reduced the number of screws from 50 to four.
- *Automation.* The company spent $3 million to upgrade facilities at its Rosevale, California, plant. A workforce of 22 initially built 300 terminals a shift, and production mounted to 25,000 a month. Careful design and engineering meant that the range of five terminals was treated as one product on the manufacturing line.

The plant cut labour and overheads by two-thirds and raw materials by half. It also increased product reliability. The most important lessons were: keep it simple and do all the little things right.

How the West lost

The division of labour

How well they implement IT depends above all on understanding how they arrived in the present position in the first place. It is necessary to go back and trace the divergent development of western and Japanese production methods. Western (indeed all) manufacturing science has been based on the division of labour. If steam was the physical force which propelled the industrial revolution, the division of labour, as Adam Smith, industrial capitalism's first and most influential philosopher, clearly identified, was its guiding principle. Batching of work to make best use of jigs operated by unskilled labour, courtesy of Eli Whitney and others in the USA, provided the next surge in productivity, obliging companies to manage stocks of work in progress as well as raw materials and finished items. By the end of the nineteenth century, the New World was at least the equal of the Old in the management of manufacturing techniques, and from that point forged steadily ahead.

The key was so-called scientific management, the contribution of Frederick Taylor, industrial engineer (and, improbably, US doubles tennis champion), and work study pioneers Lillian and Frank Gilbreth, who between them submitted labour to the same remorseless standardization previously applied to parts and processes. At the time the pioneers of scientific management, the Ford Motor Company, in effect invented the discipline of production engineering along with the Model 'T' assembly line in 1913, which produced a chassis in one hour 33 minutes instead of 12 hours 28 minutes previously. Still one of the most remarkable applications of technology to manufacturing, Ford's assembly line tripled production between 1912 (76,150 cars) and 1914 (264,972). With a cycle time of four days, including iron ore processing, it was also a brilliant example of just-in-time (JIT) scheduling. Four years later Ford factories were turning out over 2 million cars a year. In batch manufacture, division of labour gave rise to the machine shop organization: all the lathes grouped together in one part of the factory, the grinding machines in another, the boring machines in another and so on, which is still typical of, and mostly detrimental to, much of western manufacturing industry.

After World War II, which brought several further developments aiding manufacturing productivity and quality (work simplification and value engineering, for instance), the history of manufacturing development divides into two. Although it would not be evident for two decades, it was in these years that the platform for the growing influence of the newer, eastern stream was laid. As the emphasis switched from the wartime imperative of production at all costs to efficient usage of scarce resources, the Japanese were forced by post-war shortages of capital, space, raw materials and skilled labour to redesign their production machine from the ground up.

The western school, already dominated by the USA, was blissfully unaffected by such constraints. Consumers were buying everything manufacturers could make. In a sellers' market, they measured their performance on short term, internal and largely financial criteria (that is, comparing it with itself rather than with external measures of competitiveness). Such changes as they did make were grafted piecemeal on to existing organizations, factories and machines.

Unfocused factories

The result of the western tendency was factories which, like Topsy, 'just growed', without the benefit of planning, focus or rationalization. They were organizations in which undeniable busyness concealed increasing specialization, complexity, variability and cost. In both repetitive and batch manufacture, they were also remarkably cumbersome. Lead times stretched as parts which needed a few hours of machine time to make took weeks or months to thread their way along the maze of production routes. In the British engineering industry, it has been estimated that workpieces spend 95% of their time on the factory floor waiting around rather than being worked on. Hence the familiar syndrome of factories working at panic levels of urgency to fulfil routine levels of orders.

In spite of lead times measured in months or sometimes years, delivery performance was abysmal. Surveys carried out by the Cranfield School of Management found that in 1975 only half the sample plants could achieve the target of delivering three out of four orders on time, one in four plants delivered more orders late than on time, and 3% delivered every order late; even more alarmingly there was 'no significant improvement' by 1985 (BIM, 1987). In the fragmented processes and responsibilities there was no impetus for quality, and hidden costs of repairs and rework soared; far from being integrated into the aims of the process, quality control inspectors came to be hated and feared for their interruptions of precious production time.

Above all, the imperative of 'production at all costs' caused the build-up of inventory and work in progress in buffer stores and machine shops all over the factory. Come machine breakdown, failure of suppliers to deliver or strikes, the show would go on. It did, but at a high cost. Extra space and its attendant costs were the smallest element. More serious was the panoply of additional equipment and people, all of which appeared essential to keep production going while contributing not a penny or cent of added value to the product: fork lift trucks and other materials handling equipment, expediters, progress chasers, fork truck and crane drivers, storemen and quality control inspectors. All these, plus consequential costs like damage and scrap and huge amounts of paperwork, were subsumed under the catch-all title of overhead, which rose to account for 35% of a typical manufacturing plant's costs. In the UK, the inventory that most of this overhead was supporting was worth an estimated £41 billion in 1984, with carrying costs of £10 billion (Dempsey, 1986), more than enough to pay for the modernization of the entire British manufacturing estate.

The direct labour obsession

In the face of intensifying international cost competition, western firms initially chose to look for the reasons almost everywhere except inside their own organizations. To blame were, on the one hand, manifestations of unfair trading such as oil cartels and dumping by low wage rivals, and on the other 'cultural' factors such as the Japanese work discipline, identity with the firm and respect for authority. When they did turn their attention to the organization itself, western companies were handicapped by equally inefficient targeting. To cut costs, in obedience to traditional accountancy they chose to focus almost exclusively on substituting capital for direct labour in the shape of advanced NC and then computer controlled machines on the factory floor. In cutting direct labour manufacturers were extremely successful, in the sense that labour costs rapidly became an insignificant proportion of the total. But the exercise became one of steeply diminishing returns. As labour costs approach single figure percentages, further gains obviously become harder to achieve and in any case decline in significance, in both absolute and relative terms, compared with overheads and materials. Remarkably, even with the advent of CAD, the indirect areas of the firm have shown almost no productivity growth, despite the vast expansion of departmental computing. The strategic failure of firms to exploit IT in the non-manufacturing areas reflects the dismal experience of the economy as a whole. According to the US National Bureau of Statistics, productivity in the non-manufacturing sector of the economy, which now employs 80% of the total workforce, has remained essentially stagnant in a decade when the sector's IT spending has *tripled*. So far, the microeconomic effects of the new technology are almost entirely confined to the factory floor.

Fixated by labour costs, accountants and factory managers took the cost of overhead and materials more or less for granted. Where overheads were seen as variable, the variation was assumed to be mainly related to direct labour and the costs absorbed through artifical labour rates.

Attempts to deal with the much greater costs tied up in overhead and materials, not to mention poor quality, were less rigorous, tending to aim at mechanizing or computerizing existing processes, however haphazardly planned and laid out, rather than rationalizing them. Hence the proliferation of mechanical handling devices, from all manner of conveyors to automated warehouses and stores. Likewise much ingenuity in the West, particularly in the USA, was devoted to applying its superiority in computers to the factory.

One example of this was CAD, developed by Boeing, General Motors and IBM as a draughting tool in the 1960s. Perhaps more important was the development during the 1960s and 1970s of the computer-based inventory management system known as material requirements planning (MRP).

CAD and CAM

It is claimed that computer aided design (CAD), the ability to produce 3D models in a computer database, as opposed to producing 20 drawings, has been available for 20 years or more. But it was only in the latter half of the 1980s that it became more generally available by virtue of the development of desktop technology and the change in market forces within the product design field.

Computer aided design is still also available in 2D and is mostly used by companies that have simple design requirements for, say, stand-alone consumer products which do not need to integrate with any other components. Engineers benefit most from 3D, for example, when seeking to design an object which has to fit precisely into another object – like a car engine into a car body. The development of expensive modelling systems over the last few years has been astonishing but surprisingly slow, when we consider how long CAD has been around.

Increasingly, manufacturers expressed the need to be able to use CAD data as a basis for manufacturing information. Thus computer aided manufacturing (CAM) was born. It then became possible to generate 3D computer aided manufacturing data from the CAD database and supply it to specialist model-making companies, so that a definitive physical model of a product could be made, without error, in a much shorter time than would have previously been possible.

Recently, the development of a common CAD language, a software system known as IGES (initial graphics exchange specification), has allowed data from one type of CAD software system to be translated and read into another type of CAD system. IGES is far from perfect. According to the UK's Society of Motor Manufacturers and Traders, it transfers geometric information accurately but sometimes produces errors with annotation. However, the International Standards Organization is currently advancing a new translation system known as STEP (Standard for the Exchange of Product Data), which should be error free.

Meanwhile, other manufacturers are experimenting successfully with stereo lithography. This is a new technique whereby a computer controlled laser is fed with data down-loaded from the CAD system and it is able to sculpt an accurate plastic model of a designed object.

The MRP crusade

The 'MRP crusade', as it was sometimes called in the 1970s, did not in the end liberate any of manufacturing's holy places. It was an advance on the worst of the old methods, and, where it was systematically applied, provided useful results in the job-lot, batch production factories which are still the core of manufacturing industry. But, particularly in the most highly advanced and comprehensive form known as MRP II, the diseconomies of complexity set in.

MRP II, a complex closed loop system run on powerful computers, is expensive and fragile in operation, depending crucially on being fed regularly updated and accurate figures on a wide range of parameters. It takes no account of the issue of quality, or, for that matter, of the way lead times through processes vary according to the capacity available. A single-minded implementation has thrown up a few MRP II success stories; sceptics doubt the optimistic claims made for it. Widely held industry estimates are that up to 80% of MRP II installations fail to live up to initial expectations of business benefit, even if they do generally provide a means to obtain accurate operational data.

MRP I and MRP II

Materials requirement planning (MRP) is another system which was developed about 20 years ago. In its simplest form it aims to control delivery dates of goods by a computerized planning and control system. In other words, it should balance supply with demand, thus eliminating waste of time and money within the production process.

MRP II is a recent expansion of the MRP principle into a far more comprehensive monitoring and planning – an entire business control system in fact. It depends upon the integration of all the business and production aspects of manufacturing.

The successful adoption of MRP II relies heavily upon a sound organizational structure within a company capable of feeding data constantly and efficiently into a central control system.

MRP II is marketed as a system which will improve materials flow performance, allow management greater time for planning, remove functional boundaries and eliminate waste. In practice it will do none of these things if the structure of the organization it supposedly serves is not well organized, disciplined and trained.

It is a prime example of technology misunderstood by management in that most MRP failures, according to the many surveys conducted over recent years, are due to poor interface between the system and the people.

The Japanese approach

By the early 1980s it was clear that western manufacturers over a whole range of industries could no longer rely on their traditional remedies of pruning direct labour costs and computerizing production controls to achieve cost competitiveness with the Far East. The technological fix was not working. It took several more years for them to recognize the corollary, that the secret of the Japanese success was nothing to do with cheap labour (at least in the large firms, Japanese wage costs are now on a par with those of the West), not much more to do with computers and everything to do with the fundamentally different approach to the disciplines of manufacturing with which Japanese industry grew up after World War II.

Frugal methods

The reasons are complex and varied, but with hindsight it is clear that in rebuilding its industry in the 1950s and 1960s Japan did an exemplary job of going back to first principles. Not all of this was the product of Machiavellian foresight. The utter destruction of pre-war industries meant that they had to be reconstructed from scratch, initially with limited means. The need to import virtually all energy and raw materials encouraged frugal attitudes to production which were in strong contrast to the profligate manufacturing methods of the West. The lack of skilled manpower and the need to convert a predominantly rural workforce were powerful incentives to keep things simple. The fivefold rise in the price of oil between 1970 and 1974 confirmed the importance of the good housekeeping habits. Initially lacking the computer back-up of the USA and Europe (there is still a critical shortage of software engineers), Japanese manufacturers have naturally tended to prefer simple systems solutions. And for an industry which was trying to break into world export markets the importance of quality rapidly came to be paramount.

The quality gospel

Ironically, the men who built the Japanese giants took as their foundation some lessons of impeccably western origin. Most notably, they adopted the ideas on quality of W.E. Deming and J. M. Juran which are only now being put more widely into effect in the West. Richard J. Schonberger (1982) has ably chronicled the growth of the total quality control (TQC) movement in Japan after the war. Japanese manufacturers applied simple but effective principles aimed at building in quality rather than inspecting it afterwards (fault prevention rather than cure), taking responsibility for quality out of the hands of the quality control (QC) department and giving it to the line worker, making the goal one of continual quality improvement and measuring the progress with techniques such as statistical process control. By dint of such rigorous methods, many Japanese manufacturers have reached the stage of measuring defects in parts per million rather than the percentages normal in the West.

Just-in-time

Total quality control found its natural ally in just-in-time (JIT) or continuous flow production methods. Far from being merely a method of inventory control or supplier management, as it is sometimes represented, JIT is a manufacturing philosophy which drives a chain reaction of improvements in productivity, quality and worker motivation. It is simple in operation, as it needs to be in order to be understood where it is applied (on the factory floor), inexpensive and needs little in the way of computer back-up. It puts the spiral of expense, complexity and fragmentation, which in the past has driven the western manufacturing model, into reverse.

Conceptually, JIT is so simple that to some western minds it appears simplistic. Nevertheless, simplicity is a major part of its effectiveness. Not only do its dynamics attack at source the major cost components of materials and overhead, but by streamlining the production process, identifying bottlenecks, building in quality and clarifying the information needs it prepares the way for step-by-step automation. And that is exactly how the best Japanese firms have used it. Not for them the direct leap to CIM, with all the expense, radical technological and human risk that entails, but an incremental progression from one stage to the next: the reorganization of production for continuous flow dictated by the principle of job-lots of one; the slashing of set-up times with the creative use and modification of CNC tools; the creation of minifactories within the plant; and as experience grows, the linkage of the cells into larger islands of automation, and even complete focused factories. Once again, it is the organization of the information loops which is important, not the information carrier, whether electronic or not.

Just-in-time

At it simplest, JIT, sometimes called continuous flow manufacturing (CFM) or short cycle manufacturing, is a means of removing inventory by cutting batch sizes and machine set-up times. But it is much more than an inventory

continues

continued

control system. It is a philosophy of organization which affects every aspect of a manufacturing business. It works like this.

- The aim, expressed in the name, is to produce parts just in time to be used in the next stage of assembly or processing, and completed products just in time to meet an order. Ideally, the actual order would trigger the whole process, requisitioning materials and 'pulling' parts through to final assembly and dispatch. That is hard to achieve (not least because it depends on short supply chains), but the principle is clear enough: the nearer the lot size is to one, the fewer the parts and subassemblies that will be lying around incurring costs and damage and cluttering up space, since each will by definition be undergoing processing almost all the time.
- By removing the slack in the system, JIT instantly fingers weak points that high inventory conceals. If the lot size is one, a wrong part or a part made to the wrong dimensions brings the whole chain to a halt. Instant feedback is essential but also feasible, since problems stick out like a sore thumb. Management is by sight. The quicker the feedback, the fewer the faulty parts and the less scrap and rework. Productivity therefore improves.
- Quality improves, partly for the same reasons of swift feedback. There are also strong motivational effects. By clearly revealing cause and effect, JIT turns every work centre into another's customer, reinforcing good performance and correcting bad. This is a powerful influence in its own right.
- The process is dynamic, not static. As the logistics pipeline is slimmed, each bottleneck or quality problem resolved reveals another one behind it. Because cases are visibly linked to effects, problems become opportunities. The secret of JIT, if that is what it can be called, is that it triggers a process of continuous improvement.
- JIT makes manufacturing problems visible by encouraging simpler factory layouts. In the same way, it simplifies and clarifies management responsibilities. It puts control and responsibility for manufacturing quality squarely on the operators and foremen who are best placed to influence it. The information loop remains short. This not only means that routine production problems are handled quickly. It also frees managers from the fire-fighting duties, allowing them to concentrate on longer-term and strategic issues. The system decides what happens next; the manager's job is to think how to make the system better.
- JIT's virtuous circle includes market performance. The built-in tendency towards continuous self-improvement results in cheaper, better goods, often increasing market share. Since JIT dramatically cuts lead times, it also increases market responsiveness; another marketing plus. With simple elegance, JIT simultaneously encourages higher volumes and greater flexibility.

By throwing a new light on production, JIT has transformed whole industries. Inherent in its application is the attack on the largest areas of waste and the spotlighting of production bottlenecks. According to one authority at IBM (Warner, 1987), one of the heaviest spenders on CIM, 'Of all the aspects of IBM's investment in manufacturing... the least expensive, continuous flow manufacturing, is the most significant.' If it is so revolutionary and so simple, why doesn't everyone do it? There are two answers, a general and a particular. The general one is that it takes a willingness to think the unthinkable and confront issues such as machine set-up times which are usually ignored. Set-up has always been regarded as a manufacturing given, legitimized in apparently scientific measurements like economic order quantities (EOQs) which in their turn seem to justify large batch sizes. But with shorter set-up times, EOQs fall.

Toyota began cutting set-up times in the late 1960s. A first campaign reduced set-up for a 1000 ton press from four to one and a half hours. The next aimed to take it down to three minutes. The aim was SMED (single-minute exchange of die) or OTED (one-touch exchange of die) (Shingo, 1985). Methods used range from adapting existing tools to making up cheap dedicated ones. Machine tool manufacturers have now approached this problem from the other end: making all-purpose tools which can perform a variety of processes with one set-up. At bottom, the ability to engineer the time out of set-up is the platform for all the rest of JIT.

The particular answer is that JIT works best in repetitive high volume manufacture with low product variety. Not coincidentally, it is with products made by these methods, such as cars, watches, cameras, TVs, that the Japanese, who developed JIT, have made the greatest inroads. It works less well, and the Japanese have been less successful, in situations such as medium- or low-volume production of a large range of products, where long procurement lead times and demand variability may require inventory build-up and thus reduce the value of the methods used. Although JIT is always the goal to aim at, in such situations it needs help from some aspects of western computer-based planning systems (such as MRP) which were developed to meet just this kind of need.

The General Motors comparison

Here, then, is the heart of the difference between the two traditions of manufacture; not the technology itself, but the culture which governs its use. This accounts for the widely varying patterns of automation investment between countries already noted and similarly the vast difference in Japanese and US flexible manufacturing systems (FMS) performance described in the literature.

For the present, perhaps the most telling juxtaposition of the two cultures is provided by the contrasting performance of two General Motors auto plants. GM is the paradigm of the US high tech approach to modern manufacturing

Under this 'shock of the old' it is becoming increasingly evident that the mfortable cyclical progression of improvement of the past is no longer adeate. Rather than another round of serial improvements (production followed marketing followed by finance followed by...) to tighten up the slack in ch functional area, nothing short of an integration of constant improvements all these fronts will do. Hence the current emphasis on efficiency, quality d flexibility. The business as a whole must be tuned to run at maximum ficiency. It is total factor productivity that counts. The era of considering a siness as the sum of its discrete parts is over. The era of synergy through integtion has arrived.

ive towards integration

the new competitive climate, the key variables which make the difference etween manufacturing success and failure are not technology (potentially the me everywhere) and cheapness of labour (at 5–10% of total costs it hardly atters), but product design and process management: that is, how informaon is used. Manufacturing, it becomes apparent, is as much about collecting, andling, modifying and acting on data as about processing material, and it is e ability to leverage these data manipulating skills which is the new determiant of cost competitiveness. The capacity to use and modify data enabled IBM implement design for manufacture principles at Greenock and cut the multilicity of components used in personal computer visual display unit monitors nd power logic units to ten parts and eight screws. Superior information andling ability is why Japanese firms, according to one detailed account, get en times as many different parts out of their flexible manufacturing systems FMS) as US firms and utilize them almost twice as intensively (Jaikumar, 986). At a very different level, the use of information is allowing many companies to differentiate their offering to the customer by attaching information o it in some way, for example putting programmable logic controllers or computers on factory hardware, or improving overall service through electronic data interchange between the customer's buyers and their own production planners.

The need to integrate the many information technology functions within one firm coincides with an explosion of vendors attempting to deliver the means. It is here that for the enquiring manufacturer siege by acronym starts in earnest. Fuelled by the remarkable expansion of computer power and availability since the early 1950s, in successive waves the search for manufacturing salvation has thrown up computer aided design, computer aided manufacturing and computer aided engineering (CAD, CAM and CAE). Computer aided production planning (CAPP) and computer aided production management

management, just as Toyota is of the Japanese. It is therefore significant that productivity at GM's highly automated Hamtramck plant is reported to be one-third less than that of the ageing GM–Toyota joint-venture factory at Fremont, California. Nummi, the California factory, turns out compact Toyotas for the US market. It does so without most of the robots and computer power used at the later plants. It does, however, employ much of Toyota's own careful quality management and labour relations methods.

Getting the best of both worlds

It would be a mistake to overdo the breastbeating and *mea culpas* over the fading vigour of western manufacturing. Good European and US companies possess important design, engineering and marketing strengths; these should not be diminished by a manufacturing inferiority complex nor swamped by the overenthusiastic adoption of badly translated methods from abroad (the mixed results of implementing quality circles in the USA and Europe should be a warning here). Nor is it sensible to dismiss IT-intensive methods in their entirety. For instance, there are indications that, in suitably modified form, MRP for planning can be used in conjunction with JIT controls for the great benefit of job shop manufacturers, in a way, moreover, which the small Japanese subcontractors have yet to master. But for such advantages to pull their competitive weight they must be harnessed with the state of the art in manufacturing management, which above all means a philosophy to underpin and support the progressive use of the most appropriate, rather than the highest, manufacturing technology.

There is no indication yet that the West is running out of product creativity (although in the long term, creativity which is not supported by manufacturing skills and market share will wither). What is needed is integration of those design abilities, via well chosen information systems, with much better process skills. Already there is plenty of evidence to show what process improvement can do. Well documented cases show that good organization supported by modern technology can cut engineering design cost by 30%, lead times by 60% and work in progress (WIP) by up to 80%, while increasing yields of acceptable product up to fivefold. Other gains have included a 300% improvement in the productivity of capital equipment and a 20% reduction in labour costs.

But it is important to stress again that high technology is not itself the key to the improvements. Many reports have emphasized that investment in design for manufacture, quality management, production engineering and inventory control is 'at least as important' as advanced manufacturing technology (AMT), and that by using them companies can improve their competitiveness without radical changes in hardware. In one dramatic case, Ingersoll Engineers quotes a medium-sized company making engineering equipment in small batches which, mainly through the imposition of JIT principles and a few robots, cut throughput time from 25 to two days, raised inventory turns from five to 30, dispensed with 19 of its 24 fork lift trucks and cut inventory from £10 million to £2 million. Late deliveries, previously running at 40%, were reduced to 2% of the total. In balance sheet terms, the cost of the improvements, worth around £10 million a year to the company, was actually negative, the £6.5 million spent on the reshaped processes being more than offset by the one-time reduction of £8 million in inventory.

The initial streamlining is not an option. Not only does it help to pay for AMT: automation itself only works when it follows the strategic recasting of the business along the simple lines dictated by the new manufacturing logic, a fact which explains both the very patchy success of the first wave of western attempts to move direct to CIM without equivalent organizational improvement and the high performance of the fewer Japanese installations. Few people doubt that in the long term information technology and its derivatives will play a significant role in transforming manufacturing. But as one writer (Warner, 1987) succinctly put it, 'The issue is one of timing – a firm should forgo IT based approaches to solving production problems until it has exhausted conventional approaches, and then move forward into flexible automation.' The prerequisite for doing so is a strategy, analysed later in this book, which may be summed up as: simplify–integrate–automate.

References

Dempsey P. (1986). Breaking new ground in JIT, *Just-in-Time Manufacturing*. Berlin: IFS, Springer Verlag

IBM (1987). *Computer Integrated Manufacturing: the IBM Experience*

National Economic Development Office (1985). *Advanced Manufacturing Technology: the Impact of New Technology on Engineering Batch Production*. NEDO, Advanced Manufacturing Systems Group

New C. C. and Myers A. (1987). *Managing Manufacturing Operations in the UK, 1975–1985*. Cranfield Institute/British Institute of Management

Owen T. (1987). *Robots out of Wonderland: How to Use Robots in the Age of CIM*. Cranfield Press

Schonberger R. J. (1982). *Japanese Manufacturing Techniques*. Free Press

Shingo S. (1985). *A Revolution in Manufacturing: the SMED System*. Cambridge, Mass: Productivity Press

Wall Street Journal (1988). Factory of the future becomes a vision of the past. 1 September

Warner T. N. (1987). (Quoted in) Information technology as a competitive burden. *Sloan Management Review*, Fall

Towards
philosop

As the spread of the new technologies anc
other, they betoken changes of kind rather t
as it were, which is taking manufacturing i
the measuring instruments it has relied on i
volatile environment, firms could adapt anc
sequential enhancements of each function i
effort of the war years (for many firms still t
ufacturing) was followed by the era of mark
petition switched from making to selling. Tl
the 1960s favoured the view that financial k
management asset for the ambitious corpora
in boardrooms. In the 1970s the emphasis shi
the golden age of the personnel manager, the
tions for profit.

The energy-triggered recession of the l
renewed emphasis on operating efficiencies
sides of the Atlantic to shed some of the exce
preceding years of plenty. Necessary as the bl
was short term and panic led, a tactical crash
lasting strategy for growth and prosperity. Fo
uinely leaner and fitter, there was another tha
timid. This does not promise a healthy future.
consultants have been forced to take a fresh l
duction, an old idea in a new form has been gai
facturing not as a cost centre but as a source of

The d

(CAPM) developed at the same time. Meanwhile, numerically controlled (NC) gave way to computer numerically controlled (CNC) machine tools, which, commanded perhaps by DNC (direct numerical control) on MAP (manufacturing automation protocol) networks, combine with handling mechanisms and programmable logic controllers (PLCs) to produce FMS (either flexible machining or, if incorporating assembly, flexible manufacturing systems) using advanced manufacturing technology (AMT). To make matters worse, one must also fight off 'siege by choice': an astonishing number of technology vendors plying their wares.

The most ambitious in this gallimaufry of impressive sounding initials is CIM (computer integrated manufacturing). CIM is shorthand for the attempt to run an entire manufacturing business by computer. It presupposes a high degree of automation, and that every department – sales, marketing, finance, human resources, design and production – should be linked interactively by computer with the same database and with each other. In a firm run by CIM, manufacturing would be fully integrated with the rest of the functions, from automated receipt of a customer's enquiry, through the issue of the specification and order intake to manufacturing and final shipment. CIM subsumes all the rest of the individual computer aided functions. CIM is to CAD, CAM and the rest what in the human body the brain and spinal cord are to the specialized nerves of the fingers or the leg.

CIM and CIB

In their purest forms, computer integrated manufacturing (CIM) refers to the integration of the aspects of manufacturing that take place within the factory environment and computer integrated business (CIB) refers to the integration of all business activities, including those such as sales and marketing, which operate outside the factory environment. In reality, the distinctions have become blurred and the underlying management control philosophies, that is, integration of the functions, systems and human resources, have become muddled by the concern over the integration of the technology.

The issue of interface between different computer systems and software has dominated debates on the subject in western management literature and great efforts have been made in various quarters to develop protocols for exchanging information between the various hardware and software. However, experts suspect that western management is losing sight of the pure philosophies of CIM and CIB which need to underpin securely any successful technology interface.

Is integration the answer?

After a period in which manufacturers eagerly accepted that all technology was good and the more the better, a much needed reassessment is in full swing. As complicated technological solutions fail to deliver their promises, some users and commentators are questioning the utility of high technology in the factory at all. They point to the manufacturing success of companies such as Toyota, using conventional but much modified machine tools in concert with simple *kanban* control information, and advocate simplified or focused manufacturing which aims to get better results from improved organization and discipline rather than advanced technology.

Kanban is the Japanese word for record card – a simple piece of card which gives instructions for each part of a process. It was developed by Toyota in Japan as a means of controlling a just-in-time system. It is not a computer-reliant system, it is a manual system of planning and control. C.D. J. Waters in his book *An Introduction to Operations Management* (1991) describes the way a typical *kanban* system works:

> 'A more usual *kanban* system... uses two distinct types of card: a production *kanban* and a movement *kanban*. Then the process is:
>
> - When a workstation needs more materials a movement *kanban* is put on an empty container, which gives permission to take it to the area where work in progress is kept.
> - A full container is then found, which will have a production *kanban* attached.
> - The production *kanban* is removed and put on a post. This gives permission for the preceding workstation to produce enough to replace the container of materials.
> - A movement *kanban* is put on the full container, giving permission to take it back to the workstation.'
>
> The process is summarized in Figure 2.1.

The sceptics claim that using IT to control complexity and uncertainty in manufacturing plants is wasteful and misguided. Since complexity and uncertainty (inefficiently designed products containing components from too many suppliers made on highly variable systems) are self-induced, a far better solution is to do away with them by better product and process design, thereby dispensing with the need for computers at all. Thus, where companies have automated, much of the efficiency gain has come not from automation itself but from design for manufacturability; group technology may be as good a solution for a job shop's scheduling problems as computers. Most of the benefits of sophisticated flexible manufacturing cells can be gained with properly grouped conventional technology, single-minute or one-touch set-up times and well motivated staff; and JIT can obviate the need for complicated control

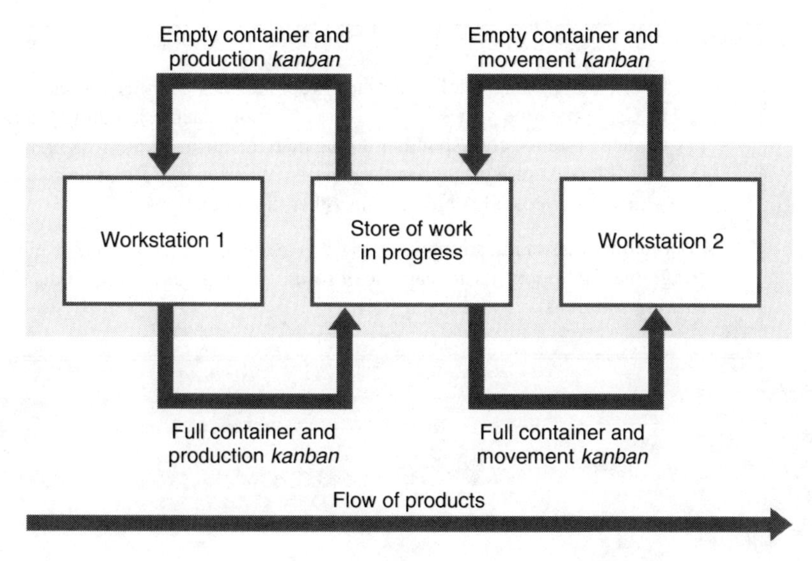

Figure 2.1 Outline of a *kanban* system with two workstations. *Source:* Waters, C. D. J. (1991).

packages based on MRP II. All these strategies show competitive use of information systems, but not necessarily based on IT. As one writer put it,

> 'The tragedy is that an IT approach treats these [complex] conditions as immutable, whereas in many cases they are not. Rather than reducing waste, an IT approach adds to it by burdening an already inefficient system with the cost of computation.'
>
> (Warner, 1987)

IT in the factory

Defining information technology is a problem in itself. It is useful to think first of the four functions of manufacturing, as shown in Figure 2.2. Using this model, IT in the factory breaks down into four categories.

(1) CAD, CAPP and CAE corresponding to product and process design.
(2) CAPM and MRP (and/or *kanban*) corresponding to inventory management, production planning and control.

continues

continued

(3) CAM, comprising intelligent shop floor devices such as robots, NC PLCs, machine tools, FMS and automated handling and storage systems, corresponding to the manufacturing process itself.

(4) The information systems at the centre, both hardware and software, which integrate and coordinate the other three functions.

Supported by communications links to other parts of the firm (such as marketing), these components are capable of forming computer integrated manufacturing solutions.

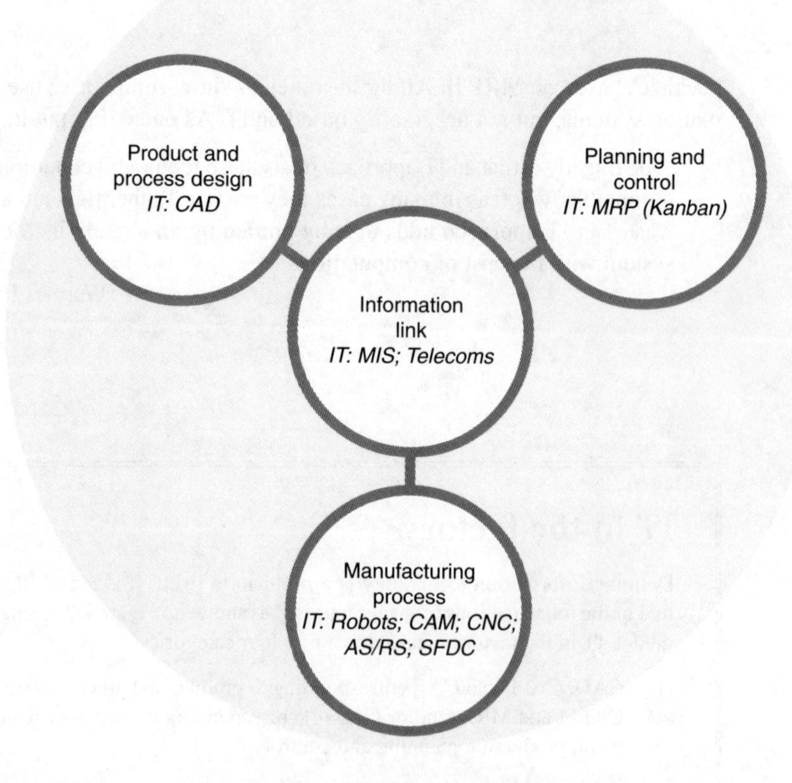

Figure 2.2 IT in the factory. *Source:* Gunn (1987).

The reappraisal is welcome and overdue. There is much truth in the saying (currently terrifying computer makers) that for most manufacturers the best way to get a high return on investment in a CIM project is to go through all the preliminary motions in the organization and then not install the computers. But if past faith in technology was excessive and naive, it would be equally unwise to overdo the anti-IT backlash. A strategic view would look to the longer term, when the company has exhausted the conventional technology means of removing waste in indirects, inventory, space, material and other overheads. It would then weigh carefully the real strengths of factory IT, notably to:

- permit effective control and monitoring of complex processes;
- reduce the cost of controlling and recording transactions;
- communicate data quickly and accurately; and
- put managers in closer productive contact with suppliers and customers.

Taking all these things together it would conclude, rightly, not only that in the long term there is no inherent contradiction between frugal manufacturing and IT, but that when both are appropriate, they support each other. The issues, as they always have been, are when to install IT and how much the organization can usefully digest at one time. Computers are not, however, an alternative to clean organization, simplified products and close relationships with suppliers and customers. At best they can enhance those benefits. So for the moment there is no doubt where first priorities lie: organization comes first, technology second. As Warner (1987) summed up:

> 'Information technology will play a key role in transforming manufacturing. But not now, not for most firms. For them it is the hard road of conventional process improvement and production system organisation that will lead to manufacturing competitiveness.'

The cautionary case of CIM

The idea of integration is elegant and simple. It describes the condition that a one or two person business naturally started out with and what the bloated, fragmented hierarchies of big companies now need to reinvent. CIM sounds like a concept which has arrived right on cue. But beware of confusing the two. Integration is organization, the means of structuring a manufacturing business to survive today's competition, and the essential precondition for using technology, including computers where appropriate, to amplify integration's effects. Integrated manufacture does not necessarily entail computerization or automation. It simplifies the business round a central information directory, accessible to everyone like a telephone list, covering where information is and who owns it. The point of integration is that the organization 'gets to know what it knows'. The directory could well be on a computer, but it does not have to be. In a small business it is typically in the heads of one or two entrepreneurs. In CIM, the firm's central information highway is by definition computer run.

Why the caution? Like CAD, CAM and the other manifestations of the computer in manufacturing, CIM is largely a technology-pushed rather than a customer-pulled solution. This is not to belittle the enormous potential of IT-based systems in manufacturing. But recent corporate history is littered with the casualties of technological overenthusiasm, and if any company aiming to install some kind of CIM is to get value for money, it will need to have a very clear idea of the great practical difficulties of implementing it.

The problems of implementation

There are three main problems, all of them the result of the businessperson's penchant for practical experiment, leaving the theorizing for later. Sometimes experimentation is sensible: some companies gained important drawing office productivity benefits from CAD as a stand-alone island of automation, for instance, before it was linked with CAM and CAE. (Others gained none at all.) More often the enthusiasm for the quick fix creates huge problems later when the faults of the overall model become apparent.

Machine integration

At the simplest level, there is the problem of integrating hardware. A substantial manufacturing company will usually have several different kinds of information processing capability for its office functions (purchasing, sales, accounting, payroll), manufacturing and R&D. Secretaries will have word processors. Some managers will be using personal computers (PCs). Almost certainly, these items of hardware will not communicate with each other in any systematic way. The central office functions will probably run off the same general purpose computer; but it is often found that many companies locate their computing resource and expertise within a financial function, and its service to the engineering and manufacturing functions is generally poor. Even assuming the will, technically the administrative machines will not be able to talk to the shop floor or the drawing office. The manager's PC may well be a stand-alone personal productivity tool, unconnected to official company data.

For most companies, the practical difficulties of hooking together this heterogeneous tangle of different makes, sizes and categories of computer into a single network using a single database are equivalent to those of a home handyman devising and installing a sophisticated telephone network and switchboard. Customizing networks for multivendor, multiapplication and multisite installations is a vast undertaking which is fraught with difficulty. Given large amounts of time, money and help from systems integration experts, it is usually technically possible. But the costs of replacing incompatible equipment are likely to be high. The intangible costs of unlearning old habits and retraining for new ones may be even higher. Such systems are inherently complex, prone to disruption and difficult to maintain. For many companies, the opportunity costs are simply too large to justify.

Integrating business functions

If a firm does decide to go ahead with electronic integration, there are greater non-mechanical problems in store. The first is at the level of the individual business functions. As computerization has affected each functional area in turn, the tendency has been to add to or speed up existing methods rather than rethink them from scratch. It is difficult voluntarily to forswear added functionality, even when it is not strictly needed. The result has been one of the great paradoxes of the modern business world: ever faster and more sophisticated hardware churning harder and harder to carry out tasks that probably should not be done at all.

Information handling tasks like this which add no value are present in practically every department of the average company. The most striking examples of the misplaced use of information are in production. Managers, machinists and warehouse staff 'know' how individual factories and parts of them operate as a mixture of intuition and custom, but even the cleverest and most intuitive have been defeated as the complexity of products, processes, layouts and handling systems has increased. Typically, manufacturing has been assumed to be an inherently complicated task, the uncertainties and variables increasing with each new product or process innovation. Remarkably, until now there has been no such thing as an overriding general theory of the plant in technical terms.

Solutions for the problems of complexity and uncertainty have created their own difficulties. One approach has been to break down the processes into understandable chunks. Each machine or group of machines has its own schedule, its own paperwork and its own stock of parts waiting to be operated on or conveyed to the next station. That makes individual processes manageable but fragments the overall system. To compensate, Western managers have attempted to bring this cumbersome factory up to date by using information technology to keep track of increasingly complicated processes and the parts passing through them. 'Stop worrying and instal something', as Wickham Skinner (1988) described the knee-jerk reaction. Sales staff have certainly sold a lot of expensive scheduling hardware and software to anxious plant managers in the cause of speeding up manufacture. But the results of their efforts have been extremely mixed. IT on the shop floor is usually complex, fallible and above all subject to human foible. In any case, the organizational consequences of IT installations are far more difficult and costly to get right than the technology itself. Few companies have yet been able to manage the huge change in job role and interpersonal behaviour required for even a simple technology like electronic mail.

In short, there seems little reason to modify a 1985 finding that

> 'Very few companies have what might be described as compelling shop window [CIM] installations running at a level sufficient to stimulate interest and application. And many other systems, parts of which may be operating under computer control, do not live up to the integrated manufacture requirements of simplicity and lower production cost.'
>
> (Ingersoll, 1985)

These damning verdicts amply make the point that automation, like IT, does not turn a poor factory into a good one. It just loads them with one more overhead. Computers exchange information fast. Used properly, they are a means to the end of making a manufacturing business profitable, not an end in themselves. It has taken the development of much less profligate methods, in particular Japanese JIT, bottlenecking and quality techniques, to focus manufacturing attention away from technology and back on basics; to understand, for instance, that the factory as a whole can only run as fast as its slowest machining centre or assembly station that causes a bottleneck.

Integrating the business

The central conclusion is that simply automating a factory's information handling traditions is pointless. A robot which cuts costs for an individual process but doubles work in progress is useless or worse. The same arguments apply *a fortiori* to the organization as a whole. If it is so hard to shake off old habits in one part of the business, how much more difficult to overcome the inertia, bureaucracy, 'not invented here', empire building and petty jealousies of entire companies. Organizational models characteristically lag well behind business and technology. The structure of many manufacturing organizations still owes more to Henri Fayol (functional structures) and Alfred Sloan (divisions) than to modern technology, just as their production lines remain indebted to Henry Ford. Such structures are characteristically rigid, hierarchical and compartmentalized, the exact opposite of the qualities needed for flexible, integrated production.

Designing a product for manufacture (still a radical departure for an astonishing number of firms) and easy service demands that R&D, engineering, manufacturing and quality control people are equally and simultaneously involved with the development of a new line from the start. Short-cycle manufacturing mandates instant feedback to suppliers on quality problems, just as it does within the firm, calling for new relationships of trust and permanence with sourcing companies. Purchasing departments must therefore work as closely with suppliers as with their own manufacturing and engineering people. Supplier firms must be willing to become more and more tightly integrated into the much larger systems of their customers, as in the case of the large UK food retailers like Marks & Spencer, which together with their suppliers in effect form a single just-in-time manufacturing and distribution system from the lettuce or tomato in the greenhouse to the sandwich in the shopper's basket.

Few companies have yet come to terms with the fact that the organizational effects of using IT in its various forms in the factory are the key to success, not the hardware or the software. Even fewer have drawn the logical conclusion that an adaptive organization, built round a responsive information system, can be a real source of competitive strength rather than a passive non-performer. Those that have done so – for example, computer companies such as Hewlett Packard, ICL and Olivetti – are, not surprisingly, in high tech industries where rapid rates of change form the normal corporate habitat. But even where the notion of information as structure is recognized, it is easy to put in place a system which hinders rather than enables.

It is possible for a corporate information system to decentralize, devolve and do away with bureaucracy. It is also possible, particularly where top management is frightened of technology and leaves the specifying to the data processing (DP) department or buys in readymade solutions from a computer maker, to be left with a centralized monster which increases rather than diminishes bureaucratic controls. British Telecom has spent years building a massive central customer database giving billing and equipment details of every account. The issue is not whether the system, reputedly the biggest software project in Europe, finally delivers the specified goods. The point to note is that in effect it requires a centralized structure for the UK telecom network rather than the network of regional operating companies, as in the USA, which other-wise qualifies as an alternative. Yet there is nothing inherent in customer billing information that makes a centralized database essential.

A manufacturing turning point: integrate or liquidate

The evidence is therefore mounting that manufacturing has reached a turning point: the most important since Henry Ford set in motion the mass production lines which have been running more or less unchanged ever since. For 30 years western companies have assumed that ever faster, more complex computer technology was the key to manufacturing efficiency. CIM is the perfect example: the big money, bells and whistles approach which has effectively made computers in manufacturing the preserve of the very large, very rich firms. It is time to dethrone technology from the pedestal it has occupied with such unhappy results since the beginning of the computer age. The sorcerer's apprentice has to be called to heel.

Now progressive integration, comprising proven techniques like just-in-time and quality disciplines, provides the opportunity, and information technology some powerful and evolving tools, to dismantle the inflexible machine shops and fixed production lines and instal the flexible machines and cells which allow rapid response and at the ultimate fuse mass and niche marketing. The reverse side of the opportunity, of course, is the challenge. If all around are tearing up the old rule books and writing new, improved ones, it is folly not to follow suit. The sure way to get it wrong is to do nothing. To paraphrase the catchphrase quoted at the beginning of this chapter: integrate or liquidate.

The rest of this book develops the implications of this central thesis for manufacturing firms and offers some lessons along the way.

AMT

Advanced manufacturing technology (AMT) is a term that no one seems able to define, although all experts seem to agree that AMT needs managing. A UK government-sponsored report *New Opportunities in Manufacturing: The Management of Technology* (Advisory Council on Applied Research and Development, 1983) defined AMT as the following:

> 'AMT is regarded as any new technique which, when adopted, is likely to require a change, not only in manufacturing practice, but also in management systems and the manufacturer's approach to the design and production engineering of the product.'

Therefore, we can assume that any new technology which threatens a radical departure from previous practices within the workplace can be regarded as AMT. All the available literature on the topic seem to be concerned with how a business deals with, in organizational and human terms, the arrival of AMT. In other words, it appears to be a 'catch-all' acronym, useful for management technique discussions.

References

Advisory Council on Applied Research and Development (ACARD) (1983). *New Opportunities in Manufacturing: the Management of Technology.* London: HMSO

Gunn T. G. (1987). *Manufacturing for Competitive Advantage.* Ballinger Publishing Co.

Ingersoll Engineers (1985). *Integrated Manufacture.* IFS, Springer Verlag

Jaikumar R. (1986). Post-industrial manufacturing. *Harvard Business Review*, November–December

Skinner W. (1988). What matters to manufacturing. *Harvard Business Review*, January–February

Warner T. N. (1987). Information technology as a competitive burden. *Sloan Management Review*, Fall

Waters C. D. J. (1991). *An Introduction to Operations Management.* Wokingham: Addison-Wesley

3

Analysing your needs and developing a strategy

No one can afford to assume that installing a computer system will automatically result in accrued benefits. Analysis and planning are the keys to any IT strategy, large or small.

There are three phases to analysing an organization's needs:

(1) Understanding the overall needs of the business.
(2) Understanding the manufacturing needs.
(3) Understanding the technology available.

After the 'needs' are determined, then a cost analysis can take place, but more about that later.

The challenge

For many companies, the urgent manufacturing challenge is nothing to do with computers or IT. Bluntly, the problem is the appearance of competitor companies that do not obey the rules under which the industry has traditionally played. They may manufacture at 20–40% less cost, offer virtually 100% quality and operate on lead times of one-third or even one-tenth of the previous norm. The problem defines the solution. The firm must cut costs, raise quality and introduce new and better products on a drastically shortened lead time. Thomas Gunn (1987) defines grade A world-class manufacturing performance as 80–100 inventory turns a year, defective parts of fewer than 200 per million and using over half the total manufacturing lead time for adding value. This compares with typical current achievements of 2–5 inventory turns, defective parts measured per cent and product times only one-tenth of overall lead times.

Even if competition is not yet at such exalted levels, the accessibility of continuous flow and quality manufacturing principles means that it soon could be. In addition, therefore, if only for protective reasons, the company must organize itself to respond rapidly to sudden and unexpected competitive changes, including shrinking product life-cycles.

Contents

The first essential in putting IT to work is to define what the manufacturing strategy is aiming to achieve within the business context as a whole, preferably in a broad formal review which serves both as a historical snapshot and, regularly updated, a reference point for the future. A typical outline strategy might commence:

> 'Our vision for the XYZ manufacturing business of the future is one which achieves world-class levels of manufacturing performance and efficiency for its products. The business must continue to increase its competitive strength and profitability by satisfying customers' needs for quality, price and delivery. Key components of this vision are as follows.
>
> - Achieve "least cost" manufacture;
> - Be highly responsive, using JIT principles to minimize activities which do not add value;
> - Be capable of manufacturing both high- and low-volume products efficiently while accommodating increased product and packaging variety;
> - Focus technology on achieving real business performance improvement;
> - Be less labour intensive;
> - Utilize simple techniques for the management and recording of quality and traceability;
> - Operate at the highest levels of quality for the industry.
>
> After analysis of world class competitors, industry norms and plant statistics, the improvement targets shown in the table overleaf have been agreed.
>
> Progress will be conditioned by the need to relate development of information systems to targets for improvement of business performance. The rate at which we can adopt new manufacturing methods is governed by the following.

	Current	*Within 12 months*	*Within 3 years*
Strategic measures			
Turnover per employee	£60,000	+20%	+50%
Gross margin	44%	50%	60%
Return on total assets	13%	20%	30%
Operating measures			
Customer service level	89%	95%	99%
Delivery lead time	8 weeks	4 weeks	2 weeks
Product unit cost	—	–3%	–10%
Cost of quality	—	–33%	–50%
Inventory turn	5	8	15

- The level of integration of physical facilities and information systems that it is practicable to attempt, given our physical layout and process constraints.
- The rate at which our workforce is willing and able to accommodate new technology and operating procedures.

Physical facilities will be simplified and integrated before major investments in technology, hardware and software are made. The simplification process will be driven by just-in-time and total quality principles. The timescales, resource implications, cost, benefits and risks expected are summarized below.'

Note that the starting point is not where the firm is now. It is where the firm must get to in order to survive. Setting such targets concentrates the mind wonderfully on business priorities, encouraging managers to subject all current routines to ruthless questioning: if we were starting from scratch, would we do it this way? Only when the business and manufacturing strategies are in place can the technical issues be addressed in their proper perspective: which IT components are needed to support the manufacturing aims – less inventory, higher quality and rapid market response – and how they are to be combined.

Preparing for IT

As well as enabling the firm to monitor its production standards and set priorities for bringing them up to world levels, a manufacturing strategy is essential to avoid advanced manufacturing technology overkill. The case for planning is made still stronger by the peculiar nature of IT. Factory IT is expensive, technically difficult to install and debug (particularly when it includes custom-built software); its order of magnitude are harder and more time-consuming to implement in human and organizational terms; and it involves considerable risk as the company moves from familiar old processes and organization to new

Feasibility report and business requirements study

One consultant, P&P Corporate Systems, recommends a structured approach to examining IT requirements, possible competing solutions and the criteria on which final choices will be made:

Feasibility report

When assessing and reviewing an information strategy, the following questions should be used as a base for a feasibility study:

- Where is the company now and what is the competition doing?
- How is the company to grow and what are the implications of this?
- What do the users want and will their needs be satisfied?
- What is your internal capability to handle computer systems?

These basic questions should be supported by a review of the information flows. You should identify existing (or future) problems that systems cannot cope with (or create). You should argue constructively the need for each aspect of these systems. Highlight the user needs that are not being met, summarize how the systems will benefit and improve the company.

Business requirements study

This study expands the feasibility report and investigates the various methods and systems needed to implement a computer system. This is a long process and even for small companies can take many weeks, It really requires an internal audit of information flows – a process called flowcharting. This study should cover all possible options to achieve the final objective. This involves evaluating competing proposals according to criteria such as cost, expandability, interoperability, functionality, performance and capacity.

The technology architecture will emerge from the document and this will form the basis for an organization-wide IT strategy. Finally you will end up with a systems proposal and this must include a full implementation cost budget covering training, cabling, hardware, software, communicative costs and so on.

Source: Director's Guide to Information Management, The Director Publications, November 1990.

and untried ones. The more technologically ambitious the project the more these considerations apply.

Dazzled by the promise of technology and, it must be said, encouraged by simplistic slogans, too many companies in the past have approached IT developments the wrong way round: technology first rather than business first. Computerizing first and asking questions about the organization afterwards is a procedure akin to 'fire, ready, aim'.

Opinion is unequivocal. Exploiting IT for competitive advantage in manufacturing (as opposed to just using it) means preparing the organization for its installation and using it for its unique potential to integrate, not as an engineering turbocharger to make individual processes go faster. Once started, manufacturing improvement is a never-ending process. Real gains can no longer be won by the piecemeal initiatives of the past, only by sustained effort and conscious commitment to stated goals. These must be monitored and upgraded over time.

The first stage of preparation is to simplify existing processes to remove the traditions which do not add value.

Simplify

Less equals more

Manufacturer, Hewlett Packard, stressed that 'The root of all evil is complexity.' In most companies, the root has grown very deep. There are hundreds if not thousands of suppliers, all of whom have to be tracked and managed. The product has a high part count. Product lines have proliferated, and factories are unfocused. Conveyors and fork lift trucks are needed to carry parts long distances between processing areas and stores to hold them while they wait between processes. When it comes to IT, microprocessors, controllers and computers in the machine shop, in the stores, at reception and dispatch, at inspection points, not to mention sales and administration, multiply uncontrollably the amount of information that is actually or potentially available.

In this situation, harassed managers tend to demand still more information processing to bring the chaos under control. This is self defeating. The new philosophy of manufacturing says that less equals more. The objective is less of everything: fewer suppliers, lower part counts, more narrowly focused factories, smaller lot sizes, smaller deliveries, smaller plants, fewer stores, simpler machines, shorter distances, less buffer stock, less waste, less reporting, fewer job classifications and, not least, less complicated information systems to keep track of it all.

Pre-automation

The first round of computer controlled factories has taught companies that pre-automation is the key to IT-based manufacturing. Pre-automation means the simplification of both product and processes. For the product, pre-automation essentially involves design for manufacture, paring down the part count and the number of suppliers, and making the item easier to build, test and service. Quality starts with product design, using techniques such as quality function development (QFD). It is taken much further by process overhaul, a disastrously neglected area in the past.

Case study: Northern Telecom

Northern Telecom, a major multinational supplier of telecommunications equipment, was faced with aggressive price competition from the Far East on some of its telephone products.

It launched a major project to redesign the telephone for automated assembly. The results were dramatic: the parts count was cut by half, from 325 parts to 156, and the assembly time was cut from 23 minutes per handset to nine minutes, a 60% reduction.

However, it is worth questioning how much of the assembly time reduction was due to product rationalization rather than process overhaul. If labour content were proportional to part count, one would expect manual assembly to be achievable in no more than 11 minutes. The process automation may be contributing very little relative to its (probably) high cost.

Quite often when products are redesigned to enable robots to assemble them (for instance, by making sure that all part movement is in only one or two dimensions), they also become much easier to assemble by hand.

Because of past neglect, pre-automation procedures must aim to subject every part of the manufacturing process to a zero-budgeting exercise, a 'right-to-exist' appraisal. Each process element must justify itself in terms of value added. On the typical factory floor, a great number of the accepted components of manufacturing add cost rather than value. This goes for incoming and most other inspection, materials handling, storage and all the people who spend their time locating lost parts or jobs and speeding them through the production process. Information handling processes should be put under the spotlight just as much as material handling processes. Computerizing inventory recording, for instance, as required by computerized production management systems, generally adds hardly any value.

The new manufacturing philosophy demands that such non-productive activities be aggressively simplified out of the system, with the added benefit that at this point the need for information links between them disappears too. The ideal is continuous flow manufacture, with batches as near as possible to one, and the simplest possible signalling methods to manage the movement of materials and parts along short and undeviating paths through the plant. In Japan, coloured golf balls, simple record cards *(kanban)* or the empty bin or pallet itself are used to 'pull' replenishments of parts to work posts, rather than IT-based systems.

Because of the random manner in which most manufacturing facilities have grown up, any information systems audit is likely to throw up some surprising revelations about how the factory really functions, as opposed to how people always thought it worked. On the one hand it will reveal the existence of outdated procedures and redundant paperwork. On the other hand, the piles of paper will often contrast with a glaring lack of communication where it most matters: between departments. One of the most common failings in industry today is lack of coordination between design, manufacturing and marketing areas, resulting in products which are difficult and costly to make, hard to modify and are then sold without regard for capacity. This is the very opposite of the qualities needed in a responsive, low-cost manufacturing system.

The medium and the message

Many firms, of course, have already made some investment in stand-alone units of information technology, for example, CAD, CAPM and CNC tools. They often assume that the natural way forward is 'more of the same': further spending on hardware and software to enhance the existing facilities and to link up the islands of automation, eventually arriving at the goal of CIM by wiring up the factory floor and dispensing with paper-based communication altogether.

But with increasing experience of high tech solutions to the supposed problems of connectivity it is becoming clear that the medium is not the message. The issue is the nature of the message, not the identity of the messenger. If the message does not need recording, use the telephone, or better still, face-to-face contact. If coloured golf balls or paper work efficiently, do not replace them. For data collection on the factory floor, blackboards and wallcharts may be better than computer terminals and networks.

In any case, simpler process design often removes the requirements for sophisticated message carrying in the first place. Group technology using JIT and quick-change tooling is the conventional technology equivalent to flexible manufacturing systems (FMS), and not necessarily the worse for that. Linking pieces of equipment electronically just because it is possible is wasteful or worse, adding cost rather than value and burying processes in a cloak of spurious high tech respectability. Although the 'islands of automation' strategy has some justification, 'users should not feel that [they] have to build a bridge between each island when a daily ferry service might do instead' (Hartland-Swann, 1987).

JIT information

To prevent non-specialists being bamboozled by technological wizardry (and to bring the specialists back to essentials), it is useful to think of information, like manufacturing itself, along JIT and CFM lines. Firms auditing their pre-manufacturing office procedures for the first time are invariably dismayed by the amount of the 'invisible inventory' that they carry in these areas. Heaps of paper in engineering, sales, purchasing and administration areas are as wasteful as piles of parts queuing up for machine time on the factory floor.

Some part of the accumulation should in any case have been reduced by previous streamlining measures in the manufacturing areas, such as simplifying the physical layout and integrating production processes. Part of the attraction of JIT is precisely that it reduces work in progress to the point where it is not worth recording at all. Still more of the overload will disappear when just-in-time principles are applied to the flow of information itself. Thus, the aim will be to cut paperwork by reporting exceptions rather than routine, reduce to a minimum activities which add cost rather than value and integrate clerical processes which do not have to be performed separately.

IBM's principles for just-in-time information are:

(1) *WIP reduction.* Elimination of redundant information.
(2) *Group technology.* Break down departmental boundaries which cause bottlenecks and delays. Encourage personal communications.
(3) *Balanced/mixed production.* Encourage parallel information flows; in particular eliminate processing in series which causes delays.
(4) *Kanban.* Up-to-date information available on demand; for example, management information 'pulled' as required.
(5) *Tightly coupled logistics.* Work to reduce distance of information flow. Where this is impractical look for low-cost automation to speed transmission times.
(6) *Supplier integration.* Extension of on-line information exchange to vendors.
(7) *Zero defects.* Zero information defects through quality and clarity of information.
(8) *Management by sight.* Present information in a visually simple way. Monitor indirect processes and make visible in departments.

(9) *Multiskilled people.* Cross-train staff to operate the total process. Understand customer and supplier needs.

(10) *Focus team.* Encourage improvement teams; five heads are better than one when dealing with complex processes.

These frugal principles are a useful antidote to the tendency to measure things which do not serve any useful purpose. The issue of MRP is relevant here. In certain circumstances MRP II systems can work successfully and well. A Cranfield study found that UK companies were more enthusiastic about MRP than any other component of manufacturing technology. But as even their champions concede, such systems depend heavily on extremely accurate stock reporting and according to Cranfield have made notably little impact on depressing overall levels of throughput efficiency and delivery performance: 'In 1975 it was hoped that the widespread use of the new sophisticated planning and control tools of MRP II would improve control. It is sad to report that this has clearly not happened' (New and Myers, 1987). At the same time, they are expensive in data collection hardware, complex in software, demanding in human time and effort and unresilient in operation; in other words, they contribute generously to overhead, more stingily to added value. Frustratingly, they work best, or rather there is less chance of their going wrong, in simpler environments, in which case, of course, they are not really needed. The arguments over MRP will continue, and there is no reason to think that present systems provide the last word. But it is important to approach the issue with the same distrust for complexity which should apply to all other shop floor systems. It is perhaps significant that the Japanese have largely done without MRP.

Integrate

By streamlining process organization and product design, a company not only begins to create the essential preconditions for installing advanced manufacturing technology. It also squarely assaults the waste in overhead costs and sets in motion a chain reaction of further cost-cutting and ground-preparing measures. Thus simplification logically implies the second governing principle of an information strategy: integration. The eventual aim will indeed be to link the different departments, creating a seamless, continuous flow business. The trick, and the source of competitive advantage, will lie in flexibly orchestrating the whole so that it equals more than the sum of its parts. But for most companies that day is still an extremely long way off. For them, the initial and perhaps most important benefits of integration will again be found in cost savings (this time in materials and work in progress), through lead time reduction and by higher quality.

Managing materials

Illogically, while in the normal non-integrated firm there are batteries of instruments and people for tracking the small percentage of direct labour cost, no one has overall responsibility for the largest manufacturing cost of all, materials. The cost of materials is rarely as low as 40%, is generally around 50% and in some cases can reach 70% or more of manufacturing cost. Traditional organization structures are not moulded around where costs occur. They are fitted around functional departments. Typically, departments process their work in series rather than in parallel, adding costs as they go. As a result, materials costs are heavily affected by design and product engineering, abetted by sales and marketing, production engineering, purchasing, quality control and inspection and finally by manufacturing, as it tries to figure out how to make the item on existing machines.

Design and production

The first aim of integration is to tap the benefits of design for manufacture and better purchasing by removing dividing lines between departments. It is well documented that over half a product's manufacturing costs are typically determined while it is on the drawing board, or at the concept stage, even if the actual costs are incurred later. For the same reason, the earlier a design or other fault is caught, the cheaper it is to rectify. Both of these facts of life put a premium on establishing close links between design and production at a very early stage in development. Simultaneous rather than sequential product and process engineering can have remarkable effects on overall costs, both directly and by cutting lead times. In Japan, machine tool manufacturer Yamazaki took a giant step towards abolishing the distinction between the two kinds of engineering: the engineer who designs the part using CAD is now also responsible for creating the CNC program to machine the part on the shop floor.

Suppliers and customers

By itself, this kind of integration has yielded impressive productivity gains, often at final assembly as a result of the delivery of higher quality, more accurate parts. But integration in the wider sense should not stop at the factory gates. One important effect of the JIT revolution has been to expose the limitations of the traditional notion of the free-standing, independent firm. Pressures for improved quality, responsiveness and cost performance have led firms increasingly to concentrate on their areas of strength, for example, the assembly of cars, rather than component manufacture. As a result of this tendency to specialize, companies are now buying in parts and even sub-assemblies which previously they would have made themselves. Such a division of labour has advantages to both supplier and customer. But to gain the full benefits throughout the whole system, sequences of suppliers and customers need to be integrated into responsive supply chains as closely as departments in individual firms.

The interdependence implied in JIT-oriented responsive supply obviously demands very different attitudes to those of the past. Gone is the previous assumption of the zero sum game, in which the customer's gain is the supplier's loss and vice versa. Instead of the ritual process of playing multiple suppliers off against each other in competitive tendering, vendors must be carefully cultivated for their contribution to quality and responsiveness. Integration and simple management control mandate long-term relationships with fewer, trusted suppliers, allies rather than adversaries. The rule is to treat suppliers as if they were your own manufacturing departments (and your own departments as if they were external suppliers). IT strategies must take these closer links into account. In industries where the continuous flow disciplines are most widely applied, such as cars and computers, suppliers and customers are already linked by extensive computer networks for the exchange of design, scheduling and delivery data. This is one of the few areas where the early benefits of computer integration are unequivocally apparent.

Marketing

Such links in the future will be the norm rather than the exception, carrying profound implications for the manufacturer itself as a supplier. Many firms have already had to adapt their own manufacturing processes substantially to meet tougher delivery and quality demands. In Europe, Ford has a computerized assessment system for suppliers, grading them 'preferred', 'acceptable' and 'unacceptable'. There are no prizes for guessing which category falls by the wayside as Ford cuts the 2000 names on its supplier list by half. For suppliers like these, the rigorous requirements of large customers have been as instrumental in bringing about manufacturing and marketing reorganization as competition by foreign or domestic rivals.

Poor delivery performance of any firm is the result of lack of communication between the sales department which takes the order and the manufacturing department which has to deliver it. Accurate information on lead times and close relations of trust (aspects of integration, by another name) are as important here as between the purchasing department and suppliers. They will become increasingly significant as pre-automation, followed as necessary by CAD, CAM and flexible manufacturing, begins seriously to affect product development times. In the past, firms have looked on shrinking product life-cycles with something akin to panic. In the light of flexible manufacturing, on the other hand, they should rather be welcomed as an opportunity for making life more difficult for me–too copyists. Reversing past logic, flexible manufacturing, whether IT or conventionally based, gives European and US firms, with their important creative advantages, the chance to take the initiative by further shortening product life-cycles, but on their own terms, as the consumer electronics giants have done in Japan. As well as supremely responsive manufacturing, this strategy also demands supremely responsive marketing and product design, each working very closely with the others.

Automate

Frugal automation

Pre-automation – simplification of product and process design and integration of departments, businesses and chains of customers and suppliers – is itself a powerful reshaper of businesses, whether or not it is followed by the installation of IT-based systems. The latter should not be a foregone conclusion, at least for the present. The Texas Instruments example demonstrates some of the competitive advantages of design for manufacture and process reorganization for JIT.

Case study: Texas Instruments 1970s

In the mid-1970s, Texas Instruments set up two calculator projects. One was to make a 48-function scientific calculator to compete with the market leading Hewlett Packard HP35, selling at $20 rather than HP's $100 in order to create a market among high school students. The second was to produce a basic four-function machine for $10.

The two projects took different approaches. Production of the cheaper machine was fully robotized. The scientific calculator team decided to go easy on automating the production line and concentrate instead on minimizing total cost at the pre-automation stage. It therefore set up an innovation team comprising parts vendors and product planners as well as factory automation specialists. They came up with a product which became the TI-30. It had just 17 parts, including rubber feet, manual, denim carry case and packing material. The single chip construction contained the first integrated

continues

continued

electronic on/off switch. Other innovations included living hinge plastics, hot stamped serial numbers, screwless snap construction and lens integrated with the case. With minimum automation, the TI-30 snapped together in six minutes. The product ramped up in production faster, was suitable for running more variants down the line, ran lower scrap rates and had a smaller direct labour content (4.8 minutes compared with 6.2 minutes) than its simpler, fully automated cousin selling at half the price. The TI-30 rapidly became the school standard in many countries, helping to displace HP as market leader, and even at the $20 retail price supported gross margins well over 60%. The calculator still sells well today.

TI beat the competition by aggressive innovation at the pre-automation stage, simplifying the product and integrating the engineering processes. Twenty years later, the lesson remains the same. World-class manufacturing is about organization for manufacture, not IT. Automation comes second, not first.

In Japan, the production feats of Toyota and other JIT champions were initially accomplished largely with conventional technology and self-developed machines, although on this solid platform FMS are now being built at increasing speed. In the USA, companies which rushed to implement CIM are slowing down as they learn the 'less equals more' lesson and investigate cheaper, simpler, incremental ways of gaining quality, delivery reliability and responsiveness. Frugal automation, like frugal manufacturing, is the watchword, forgoing IT-oriented solutions until the benefits of conventional technology have been exhausted, and only then progressing to flexible automation where it is clearly necessary.

Factory of the future becomes a vision of the past

Once seen as the key to stronger manufacturing, high tech machines increasingly are taking a back seat to beefed up training. 'The trend is clearly back to basics,' said John Vinyard of GenCorp, Akron, Ohio, which is now teaching all workers to be quality control inspectors.

continues

continued

Even firms that embraced automation pull back. With factories now 70 per cent paperless, TRW, Cleveland, emphasizes worker productivity. 'We have so much technology available that it's choking us,' said TRW's Arden Bement. The benefits of highly automated factories 'aren't obvious', said John Reid Jr of Standard Products, Cleveland, noting that old-fashioned techniques simplify the manufacturing process and improve communication. 'Sometimes you need the paper.' Fresh data from the National Bureau of Economic Research, Cambridge, shows that Japan trains its factory workers more than the USA does.

Source: Wall Street Journal, September 1, 1988.

The importance of approaching advanced technology, and particularly IT, in this way can hardly be overstated. Complicated and fragmented processes demand complicated and fragmented information systems. The risks are high, the investment higher and the benefits fewer. However, where it is installed in the right place and for the right reasons, AMT can have remarkably short payback times. On the one hand it attacks the largest and most vulnerable costs in overhead and materials. On the other hand, improvements in quality and market response make for great competitiveness and increased market share. There is also a strategic gain, progressively more significant although difficult to measure in financial terms, in the engineering know-how and experience which accumulates from a deliberate step-by-step approach to IT-based manufacturing technology.

Case study: Lucas Diesel Systems

Lucas Diesel Systems at Sudbury is proof that the key to bringing a mediocre manufacturing plant up to world-class standards is not expensive advanced technology but rigorous thinking, clear responsibilities and simple lines of organization. The Sudbury plant makes fuel injection equipment for diesel engines, high precision items made in large volumes. Once state of the art, by 1984 it had dropped behind competitors in quality, productivity, lead times and delivery reliability. It could count on little help from the parent, which was also in financial trouble.

continues

continued

Having realized their problems, managers' first key decision was to sit down with a blank sheet of paper. Never mind the site's sacred traditions, what must it do to give itself a future? The answers would leave no part of the operation unaffected. First, it must make real cuts in unit cost. Second, it must radically improve quality by getting it right first time. Third, it needed to enlarge its traditional notion of training to give work people the tools to do a bigger, multifunctional job. And finally it had to cut lead times, which would also begin to carve chunks out of the factory's mountains of work in progress (WIP).

To do all this, Lucas realized it would have to overhaul its entire management structure. Under the seven-tier existing structure, responsibilities were blurred. It was hard to solve problems because no one 'owned' them. No one was directly responsible for thinking about strategic issues like lead times or quality. Communications, according to one manager, were like a version of Chinese whispers.

Lucas's second key decision was to simplify by reorganizing its existing resources. It knew it could not count on large allocations of capital to spend on advanced technology. But it looked on that as an advantage. Managers believed that bootstrapping the factory from its own resources would (if it worked) have a powerful motivational effect on everyone in the plant. They also understood that by concentrating attention on essentials, a simplified and integrated structure would help to identify the strategic points where capital investment would be needed to amplify the gains in quality and responsiveness.

Sudbury's solution was to adopt JIT principles. It dismantled the old machine shops and reorganized existing production equipment into manufacturing cells – three factories within the factory, each making a family of parts. Instead of zigzagging around the different shops, parts travelled straight down the line, leaving it only to go through a large central heat treatment area which could not be moved or replicated.

The simplified physical layout was reflected in a radically simplified management structure. Each product line functioned as a separate business. Performance responsibilities were clearly established. Management layers came down from seven to three, which itself aided problem solving. Many problems simply vanished; since they were the product of faulty communications, better communications just dissolved them. Real problems were instantly evident; as was responsibility for resolving them.

Perhaps most striking in all this is the massive part played by a motivated workforce. In fact, there is little mystery about the wave of workplace creativity unleashed by the new organization, which is apparent in every plant run on similar principles. JIT gives people responsibility for tasks which they can influence. By making the purpose of each operation visible, it provides a strong stimulus for ideas to do it better.

After four years, the Sudbury plant nosed into profit. It cut lead times by three and improved quality. Its market share improved as a result. Using

continues

continued

much less space for the same output, it absorbed another line from a different factory. More important, it launched on a process of continuous improvement which has no obvious end. For example, the plant managed down its inventory levels as confidence in JIT grew in the cells. Stock turn, which moved from 22 to 34 when the factory was reorganized, doubled and doubled again as manufacturing teams gained experience of *kanban*-type controls.

The financial incentive to get it right applies with greater force to western companies which have much more slack in the business system to remove. Many companies have found that they can fund substantial advanced manufacturing technology (AMT) and IT investment out of the one-off reduction of inventory resulting from the adoption of JIT principles. In that sense, the present bloated stocks of materials, parts and finished goods cluttering European and US manufacturing plants are potentially a godsend as well as a curse: an unearned, unlooked for source of funds within existing working capital.

Ten rules for setting up a strategy

There are four steps to setting up a strategy.

(1) Define the business goals.
(2) Simplify: redesign the product for manufacture, the process for continuous flow delivery and control the existing processes.
(3) Integrate processes, departments and suppliers to support the goals of simplification.
(4) Implement the IT solution frugally and in stages, only when you are sure you need it to achieve those business goals.

IT in manufacturing is hard to get right, since it does not in the end depend on the technology but on organizational factors. Computer integration is a double-edged sword, having the potential to be either a powerful weapon in the armoury of competitive advantage or a burden of useless and costly complexity. Changing the metaphor, integration is also a Pandora's box: although simple in concept, in execution it must open up every aspect of the firm's activity to scrutiny if the benefits are to measure up to the effort. The revelations are often a nasty shock. Moreover, in many cases the expected benefits have not materialized. In the UK, New and Myers (1987) found that almost half the plants experienced no significant gains from CAD and CAM, two-thirds reported low or negative payoffs from FMS, and more than three-quarters complained of poor returns from robots – although the survey did not identify over how long a period. The moral is not that AMT is inherently ineffective; just that it has been beyond the capacities of most companies to implement successfully. On the positive side, experience yields some important rules of thumb which can be summarized as manufacturing technology's ten commandments.

Checklist

(1) *Have a plan.* The single most important rule. Without a manufacturing plan addressing a business strategy, the company is simply automating its problems or wasting its technology investment in piecemeal solutions which close off the options in other areas. The plan is not a guide for a one-off improvement programme which can then be retired to the archives, although it may contain more or less self-contained implementation phases. It is a constantly evolving document which changes to take account of both external (technological, competitive) and internal (product, process, organizational) events. Manufacturing improvement is a never-ending process.

(2) *Be aggressive and committed.* A move towards 'real time manufacture' – just-in-time information handling supporting just-in-time production – transforms every aspect of a company's operations. There is no point in undertaking the major changes involved unless you set ambitious goals. For example: cut inventories by 70%, indirect labour by 50%; reduce set-up and lead times by 75%, development time for new products by 50%; improve quality by 90%; achieve throughput efficiency of at least 75%, orders on time of 95%; all within a timescale, say, of no more than three years, which everyone involved can expect to see through. It is important to believe and stress that such improvements are achievable (as well as imperative). The new philosophy is a rigorous taskmaster. Insisting on such objectives demands total commitment from senior managements, without which the programme cannot in any case succeed.

(3) *Keep it simple.* Prefer simple solutions and systems to the complex and, until you have built up experience, proven technology to the untried. If the people who are to operate any part of the system cannot understand it, it will probably fail to work. Remember the large number of failures that result from trying to make too large a technological leap: walk, do not run.

(4) *Avoid over-integration.* This may seem paradoxical. But there is a trade-off between flexibility and integration, most visibly in software. The more highly integrated the system, the harder it becomes to modify sections of it without having to change everything else, and the more likely that failure in one part will cause failure of the whole. In other words, an integrated system is only as good as its weakest part. In practice, it is unrealistic to expect all the subsystems to be at an identical stage of development. It is therefore safer, although less elegant, to confine full integration to areas of stability (in software: communications, transaction processing and data collection), and adopt a modular approach with, perhaps, manual interfaces elsewhere.

(5) *Plan top down, but implement bottom up.* Planning needs a central, strategic view to set priorities. Implementation generally builds from the operator and machine tool upwards, as in the Japanese model, in order consciously to build experience. The accumulation of learning is a vital and ongoing part of the project.

(6) *Target cost areas where changes can have the quickest and most conspicuous impact.* Good examples are inventory or quality. Aim to get 75% of the benefits at 25% of the cost and in 25% of the time. One UK electronics manufacturer is reported to have made annual savings of £2.5 million as a result of a one-off investment of £35,000 to implement JIT (NEDO, 1987). This not only releases precious investment funds for further improvements, it is also important as a factor in proving to sceptics that radical change is both possible and worth undertaking.

(7) *Break the traditional financial justification mould.* This can seriously misdirect manufacturing investment, and has done in the past. The radical new performance possibilities in responsive manufacturing require a fresh set of accounting criteria (see Chapter 10).

(8) *Expect constant organizational change.* Except in a very few companies, organization is notoriously slow to reflect technological advance or business change. There is 100% certainty that organizational structure will alter with integration, and keep altering. To some extent, integration implies centralization. It is important not to let this planning principle cut across the equally important principle of devolved control.

(9) *Enlist a champion.* Until the notion of continuous change and improvement is absorbed into the corporate bloodstream, it needs constant reinforcement. A common element in most success stories is the presence of a visible, vocal and very senior champion to promote the technology, protect it from the diehards and provide a rallying point for the growing band of the converted.

(10) *Recognize the human factors as paramount.* An integrated plant redefines functions, jobs, skills and relationships. People must be willing partners in the changes, not guerrilla resisters. Companies that are successful with advanced technology have all enlisted the effort and commitment of their workforce, from operators to senior managers. There are no exceptions. Adequate techniques are necessary but not sufficient. Only motivated people can make them work. This requires a sustained effort in communications and training; in fact these become part of the integration process.

References

The Director Publications (1990). *Director's Guide to Information Management*. The Director Publications

Gunn T. G. (1987). *Manufacturing for Competitive Advantage*. Ballinger Publishing Co.

Hartland-Swann J. (1987). How much integration? *Industrial Computing*, February

National Economic Development Office (NEDO) (1987). *Winning with AMIE*

New C. C. and Myers A. (1987). *Managing Manufacturing Operations in the UK, 1975–1985*. Cranfield Institute/British Institute of Management

4

Getting help

Self-help and its limits

Where does a company look for help in planning and implementing a manufacturing system?

The short answer, with provisos which this chapter will explore, is: to itself. The themes running through this book are:

- The need for companies to review their manufacturing resource from the ground up in order to change it from a cost centre to a source of competitive advantage;
- The importance of information as a strategic tool in making this transformation;
- Putting manufacturing technology and IT in its rightful place as servant rather than master of the modern factory; and
- Mastering the organizational skills to get the best use from it.

It is abundantly clear that few companies can foresee what changing technology will mean for their organization, personnel and products in the long term. Rather than cast around for such non-existent certainties, the key management task at any technological stage is to realign the neglected manufacturing process with overall strategy and think through the information systems which are needed to support it.

This learning process is essential and itself part of the struggle to build competitive strength. Indeed, in a situation where there is no fixed technological goal, only the prospect of continuous improvement, learning, becomes part of the organization's mission. As Thomas Gunn (1987) pointed out:

> 'The real value of the plan lies in the experience of establishing the plan, the education and training and team building and commitment process entailed in creating it, and the uniting of the human resources of the organization behind a common goal of achieving advantage in manufacturing. As Dwight Eisenhower once said: "The plan is nothing. Planning is everything."'

In any case, outsiders who know more about your business *and* new manufacturing methods than you do are thin on the ground. Banks and financial institutions are naturally finance oriented. Despite the ambitious claims, no equipment supplier is in a position to provide all the advanced manufacturing answers. Computer makers, machine tool manufacturers, large electrical and electronics firms, software houses and consultants, even major construction firms, all offer different approaches to factory automation and information, and each has its own particular axe to grind.

This is not to suggest that all outsiders are useless. Few manufacturing companies have all the necessary manufacturing, engineering, software and systems skills in-house, and most recognize the need for occasional external assistance. The point is rather that change to secure long-term competitive advantage in manufacturing is not a one-off affair but a continuous process which can only be founded on deep, internalized awareness of the evolving needs of the manufacturing function. Competitors will eventually replicate moves to focus factories, cut set-up times and use new technology. Over time, competitive advantage will accrue to the manufacturer who builds the organization that can sustain the fastest rates of technical change within a steadfast culture of quality and commitment to the customer. This capability demands an intense commitment to internal resources: to investment in scarce manufacturing engineering skills, in training and education, in information systems and in personnel policies which will attract and retain the best people.

Questions for automation

'Is industry practically ready to properly plan, utilize and make better profit from the "most advanced automated production lines"?

'Will industry utilise the remaining workforce...fully on the new line? If a maintenance man is needed for the new automated line, is it not better to buy a semi-automatic line with one operator and train the man to be a multi-purpose worker (maintenance-operation-QC-transport), and accumulate the know-how for the next investment chance of full automation?

continues

continued

'Is it certain that the product being bought is not an off the shelf machine, which means everybody's machine (and "nobody's machine") that does not give any significant advantage against competitors?

'Has enough supervisor, QC and maintenance training been prepared, to utilise the new line well in advance?

'Is it certain that the automated line does not mean only automatic transportation or mechanical material handling?

'These should be essential questions for any investment in automated lines…. The old JIT concepts for the machines – slow speed, high reliability, flexibility, simple and easy to care for, inexpensive, full of in-house modifications, production facilities and full utilisation of the operator – are also applicable in the age of automation and robotics.'

Source: Abe, 1987.

In short, the core resource must be internal and the approach resolutely learn as you go. External assistance should be used occasionally. There is such a variety of learning aids available that companies would be foolhardy to ignore them altogether. Some of them are free (or nearly free), others not. Some are entirely external to the firm, while others, perhaps more interesting, result from the changing nature of supply relationships throughout the business chain. Consider that in streamlined, focused manufacturing a company may buy in 50–80% of its final product, often from a very few vendors. As an intermediate supplier, the firm may be linked in a similar way to its most substantial customers. Perhaps both its incoming and outgoing shipments are transported by a specialist distribution outfit. Thus, the company is affected as much by what goes on outside the factory as within. In this 'virtual factory', as it has been called, the tightening physical supply networks provide important and often neglected sources of assistance in developing IT and manufacturing capability, as well as commercial advantage.

The sources

There are six main sources of knowledge, and sometimes also financial assistance, for enquiring companies. They are:

- books and current practice,
- official support programmes,

- business customers/suppliers/distributors,
- equipment suppliers,
- professional institutes,
- consultants.

Current practice

Current practice is available in books, through factory visits and attending conferences. A selection of the most important recent English language documentation on developments in manufacturing is listed in the bibliography. The state of the art is changing fast, particularly in the USA, where the world-class manufacturing movement has built up a fine head of steam. Management journals such as the *Harvard Business Review, The Economist and Management Today* carry a steady stream of articles by academics and practitioners on new manufacturing approaches, quality, JIT and IT subjects: read them. Some specialist computing, production, quality and materials handling journals are worth consulting. Factory visits are a useful form of basic technology transfer. Advanced business partners, either vendors or customers, are obvious candidates for visits. Equipment suppliers maintain their own showcase plants which they are only too happy to show to prospective customers. The obvious provisos apply here: they *are* showcase plants, often set up to demonstrate what is technologically possible rather than what is strategically and financially advisable, and to address *their* needs (selling equipment) rather than yours. Plants in the same or even completely unrelated industries can offer valuable insights in adapting new techniques to different conditions. Competitors or parallel producers (that is, non-competitors) in the same industry are a specially fruitful source of knowledge. It is surprising how much practical information can be gleaned from monitoring a competitor's recruitment ads for systems people and other computer literate staff. Other methods include approaching the rival as a potential customer or supplier. From the other side, some companies have found that showing off their best plants yields important and unexpected benefits in staff morale. In the USA, Harley-Davidson discovered that operators were much more persuasive factory guides than corporate PR personnel and assigned one to each visiting party.

Official programmes

Spurred by the need to encourage manufacturing industry to become more competitive, governments in almost all industrialized countries run programmes favouring the early adoption of advanced manufacturing technology.

The UK is unique in having reversed its posture under the present government, moving away from direct support and national programmes to a consultancy approach.

In the early 1980s the emphasis was on a wide range of capital grants, with dedicated schemes to support (among others) CAD/CAM, robotics, flexible manufacturing systems, microprocessor applications, fibre optics and software. There was also a spread of advisory services. In 1983 the Cabinet Office's Advisory Council on Applied Research and Development (ACARD) noted that

> 'with the wide variety of schemes now in existence, a manufacturing company will find itself eligible for government support for almost any installations of manufacturing technology capable of raising its competitiveness.'

Since then the government has rationalized the previous system of investment grants in the most radical way possible: by scrapping it. The Department of Trade and Industry (DTI)'s direct support for AMT at the user level is now confined to consultancy under the DTI's Enterprise Initiative, which seeks to aid small and medium-sized independent firms (up to 500 on the payroll) with specialized consultancy over a range of business functions. Basically, the DTI will pay half (two-thirds in Assisted Areas and Urban Programme Areas) of the cost of between five and 15 man days of consultancy in manufacturing systems, quality, marketing, design and financial and information systems. A firm can qualify for consultancy support for more than one function. Under the Regional Initiative, the renamed rump of the previous regional development programme, small firms with fewer than 25 employees located in Development Areas can in addition apply for investment aid of 15% of the cost of fixed assets up to a maximum grant of £15,000 and innovation grants of 50% of the agreed project cost up to a maximum of £25,000. There is also a Research and Technology Initiative which encourages joint research between companies, education establishments and independent and public sector research bodies.

At national level, the UK has been less purposeful in supporting advanced manufacturing than its rivals, having neither the targeted defence procurement programmes of the USA and West Germany, nor the national development projects of Japan and France. The Alvey programme, now the UK's publicly funded answer to Japan's fifth generation project, has now ceased. The Science and Engineering Research Council directorate dealing with computers in manufacturing has limited funds at its disposal.

Like other European countries, the UK participates in the EC's scientific research programmes. The Esprit programme on information technology has been spending $25 million a year on manufacturing research, and is now gearing up to spend $1.3 billion on experiments with CIM. It has also earmarked $900 million for communications and $140 million on industrial technologies. Not surprisingly, larger companies with time and resources to spend on bureaucratic procedures have done best out of the European initiatives.

National programmes

The early impetus for CIM came from the US Department of Defense. Its ICAM, ECAM and Techmod programmes, developed in the wake of major alarms over the financial health of defence contractors like Lockheed and Chrysler, aimed to help its largest suppliers to cut costs, improve productivity and develop their technological base through advanced manufacturing and computing techniques. Given this support, it is not surprising that US defence contractors are the most vigorous champions of CIM. However, the government expects civil industry to benefit from the transfer of technology developed from military programmes, and anecdotal evidence suggests that this is now taking place, albeit at a high price to the US taxpayer.

The Japanese route to CIM is rather different. At the broadest level, much benefit has been derived from various research programmes set in motion by the Ministry of International Trade and Development (Miti). Typically, as with the present attempt to develop the so-called fifth-generation computer, Miti persuades firms to work together in groups during the development phase, with commercial hostilities reserved for the production stage. Miti's recent major manufacturing project, for which $130 million was earmarked up to 1992, was for experiments in 'best of both worlds' manufacturing: systems so fast and adaptable that they are equally cost effective for low volume/high variety goods and for high volume/low variety items currently produced on dedicated transfer lines.

Within companies, however, there is strong underpinning for all kinds of manufacturing technology in the very high production standards already achieved by conventional means, and the constant supply of highly educated young engineers turned out by the country's education system. Some 70,000 new engineers a year graduate in Japan, nearly ten times as many as in the UK. Interestingly, unlike westerners, the Japanese assume that flexible manufacturing, whether automated or not, offers particular advantages to small job shop factories. The constraint is not technology but the availability of skilled engineering manpower, of which Japan has more than any of its rivals. The uptake of appropriate technology is greatly aided by the historically low cost of capital in Japan and the banks' longer-term view of capital investment generally. Meanwhile, tight long-term trading relationships with vendors and customers, favouring the close interchange needed for joint development of high quality and just-in-time delivery, is encouraged by the grouping of many firms in *zaibatsu*, loose 'families' of companies, generally grouped round a bank or a trading house, which are often linked by cross shareholdings as well as long-standing commercial ties. Such groupings tend to contain or develop many of the important generic industrial disciplines. In the Furukawa group, for instance, Fanuc was a spin-off from Fujitsu, and Fujitsu, with its strategic product line of computers, telecoms and semiconductors, itself was spawned by Fuji Electric.

continues

continued

Different support models are found in other important advanced manufacturing countries such as Germany, France and Sweden. Both Germany and Sweden have obtained impressive results from FMS-type installations. (As in the case of Japan, this is entirely consistent with the underlying theme of this book: far from being a solution to incompetent manufacture, automated, IT-based manufacturing has most to offer the companies which are excellent performers already.) Sweden, perhaps because of the national influence of one of the most important robot suppliers, ASEA, is proportionally one of the largest users of robots in the world. Both Germany and Sweden use public money to support the use of advanced manufacturing technology in defence supply projects, with similar aims to the Pentagon in the USA. In Germany, the government also provides financial support for a manufacturing technology programme which aims to help users to integrate IT with development and manufacturing processes. The country is particularly strong in its network of regionally and federally supported research institutes which perform an important function for small and medium-sized firms without their own research departments. From a lower base France has evolved extremely ambitious manufacturing technology goals and is spending heavily to reach them: a variety of government grants and incentives is available to support most advanced manufacturing applications.

Source: Abe, 1987.

Customer/suppliers

Useful though such superordinate research is, for most companies the present most urgent need is not to install AMT, MAP, TOP and huge relational databases which can be accessed from anywhere within the firm, but to improve competitive performance without at the same time prejudicing future moves towards integration and flexibility. Since the short-term name for that is satisfying the customer, a good place to start is suppliers and customers or distributors. Under intense competitive pressures, manufacturing is becoming increasingly specialized. In pursuit of sustainable competitive advantage, firms are thinking harder about where their real strengths lie and where they should rely on other people. Why do in-house what experts do better outside?

While buying in or subcontracting reduces internal complexity, in theory making the factory easier to manage, it does not necessarily reduce the complexity of the system as a whole. Only if the supplier(s) can match the efficiency of the newly simplified factory will there be overall system gain. As many frustrated smaller JIT converts have found, just-in-time within the factory walls represents only a fraction of the overall potential. The implication: a complete rethink not just of manufacturing, but also of the company's supplying and purchasing roles, in both of which it will seek to encourage qualities that are probably the reverse of those which have driven its previous behaviour: partnership rather than confrontation, long-term development of joint advantage rather than short-term price gain, sharing knowledge rather than protecting it, fewer sources/outlets rather than more, highest possible quality rather than the minimum that can be got away with.

These relationships of increasing interdependence have implications for a company's information systems, and through those for its manufacturing technology.

- In the US motor industry, computer-generated component designs are not transmitted from the parts manufacturer to the assembler and accessed by either party. In the past, the auto maker drew up parts specifications in great detail, which component firms competed to match. Now minimal specifications encourage component makers to design and innovate, with easy modification by the assembler. In the UK, GKN developed a CAD program which can tailor a composite spring for a particular application in minutes rather than the previous three months. This joint manipulation of geometric data is a deepening of existing electronic links through which Ford, for instance, transmits details of weekly production schedules and daily advice of just-in-time deliveries required from its supplier community.
- Canny suppliers are deliberately tying in their customers by creative use of IT. In one well known example, US drug wholesaler Foremost McKesson provided its pharmacist customers with data entry terminals for automated reordering. The convenience of the system persuaded many clients to double and triple their orders, in many cases abandoning links with other distributors. In the UK, ICI's plant protection division provides agents and distributors with viewdata terminals which, as well as answering stock enquiries, can help sales staff to advise growers on the best product to use for particular plant diseases. Similar systems are beginning to appear in supermarkets as companies start to use information to differentiate their product for their customers.
- Marks & Spencer's close attention to the manufacturing operations of its suppliers is well known. It is sometimes said that M&S is the biggest integrated manufacturing operation in the country, even though other people own the factories. As part of the relationship, M&S provides consultancy for its vendors to make sure they meet the buyer's quality requirements. Another element in the partnership is the electronic sales and ordering links under the generic name EDI (electronic data interchange) which for major retailers such as J. Sainsbury, Boots and B&Q stretch from the laser scanner at the supermarket checkout right back to the supplier's terminals.
- Mercedes Benz developed a prototype simulator which is much like a flight simulator for training airline pilots, in that it gives an accurate sensation of

driving a car. But the point of the device is that it can simulate the different qualities of every Mercedes model, large or small, diesel or petrol engine, automatic or manual transmission. It can demonstrate the optional extras and show how the car behaves in different road conditions. Finally, it can tell the customer what his or her customized package will cost – financing, insurance, shipping – how long the car will take to make (say, 3–4 days) and process the order electronically, on the spot.

These examples emphasize the nature of manufacturing as part of a logistics process stretching from suppliers to customers, with information passing across corporate demarcation lines. The company's interests both as purchaser for its own manufacturing process and vendor to others require that it take full advantage of its partners' experience in design, manufacturing, delivery and information technology. Electronic links at the commercial level have become the norm, with, for instance, interchange of orders and invoices. The linking of design databases between suppliers and customers leads inevitably to further linking down into manufacturing itself. Some people believe that this is the major influence 'pulling' IT into the factory and it is important to bear this in mind while planning to move towards CIM. The solution may need to be driven as much by the need for slick communications with the outside world as by internal manufacturing considerations. Suppliers and customers are partners, not enemies, and as such form an integral part of the 'virtual factory'. Integrating their information needs is a formative stage in the development of a flexible, responsive business.

Equipment and software suppliers

There are many different types of equipment makers, and the list is steadily increasing. Computer companies such as Hewlett Packard, Honeywell, Bull, IBM, ICL and Unisys are moving out from their traditional home in back office dataprocessing departments into wider applications. Minicomputer manufacturers, such as DEC, Hewlett Packard and Prime, build on their experience in engineering design and shop floor applications in offering technically oriented solutions. More software oriented suppliers like Cincom, Hoskyns, McDonnell Douglas and MSA concentrate on application functionality. The common thread through all of them is the heavy emphasis they place on CIM as an important goal. Less dominant manufacturers tend to stress their ability as 'systems integrators' (that is, their prowess at getting machinery from different makers to communicate with each other).

Machine tool manufacturers approach the problem from the metal bashing or component insertion end. Just as they originally developed multipurpose machining centres from simple lathes and boring machines, then combined them in flexible machining and manufacturing systems, now they are aiming to graft still more sophisticated controls on to their hardware. The ultimate

goal: groups of general purpose machines which combine the inflexible transfer line's economies of scale with flexible manufacturing's economies of scope. Yamazaki, which has built a series of increasingly sophisticated factories to construct its huge range of machine tools, is a good example of this line of thinking. So is the Italian firm Mandelli, a notable European success story. The privately owned machine tool firm claims around 10% of the estimated 280 FMS systems currently in place in the industrialized world.

A third group of AMT suppliers comprises industrial and electronic conglomerates which have either developed automated factory systems from their own manufacturing operations or bought the components in. The standard bearer of this approach is GE, which has acquired CAD/CAM, semiconductor and robotics firms in its attempt to become the major US suppliers of factory automation. Philips and Siemens are two European contenders which believe they can combine DP skills with their expertise in making the specialized integrated circuits which will lie at the heart of factory communications. By contrast the West German heavy engineering group Mannesmann, Fiat's Comau subsidiary and Renault in France are all trying to leverage their own manufacturing skills. Software consultancies (such as Hoskyns, Logica, CGS, SD/Scicon in the UK; McDonnell Douglas and EDS in the USA), branching out from discrete programs in factory control, are also strong runners in the integration stakes. Finally, the giant US construction firm Bechtel recently set up an advanced technology division which offers to build turnkey CIM plants to a client's specified needs.

At the current stage of development, even ordinary users who have successfully integrated hardware and control software into flexible automated systems are eager to sell their know-how. None of them, however, has yet been able to impose itself exclusively, and buyers would do well to ask any supplier some searching questions.

- *Can the solution accommodate a step-by-step approach?* For most companies, certainly those without experience, step by step is better than too large a technical leap.
- *How flexible is flexible?* Western manufacturers have been much less successful in using their FMS flexibly than the Japanese. In many cases they may have been better off buying simpler dedicated tooling and modifying it to their own ends than carrying the cost penalties of supermachines whose flexibility cannot always be exploited but which must be kept occupied for financial reasons. In this way, the apparent flexibility of the machine paradoxically limits the flexibility of the system as a whole. Too little flexibility is a constraint. Too much flexibility, as too much automation, is a waste, adding complexity and uncertainty and then extra data-processing capability to sort it out again. Buyers should beware of technology for its own sake. Find the supplier which will install and support the simplest possible system for the circumstances.
- *How tied will you be to one supplier?* No supplier is best at everything, and

in a situation where the state of the art is changing fast it is not sensible for a company to lock itself exclusively into a single proprietary source. A better approach is to select a portfolio of the best solutions for each individual application and loosely integrate them. This has the advantage of maintaining system resilience. Bearing the movement towards integration in mind, it is important to make sure that multivendor systems can work together in harmony. This is the selling point of the systems integrators. In the long run, standard interconnections between machines will certainly be important. But harmony can often be achieved with much less than total (that is, expensive and complex) electronic communication and integration.

- *Can the vendor show reference sites (not just its own) for its systems?* How many? If it manufactures itself, does it use the system it is selling? How many customer sites does it support in the UK or Europe? Apart from the giants, most of the computer firms aim at 'vertical' or niche markets. Find out which are strongest in your own industry, and check out the reference sites.
- *What about quality?* This is a critical point. Suppliers need to prove the quality of their equipment. Find out whether the vendor, as well as its users, meets BSI standards (BS 5750) for its own operation; how it measures quality; whether it has quality improvement programmes of the sort you would expect to find in a world-class manufacturer; the focus of its R&D programmes. But quality is not just a matter of technical specification.
- *What is the level (and cost) of support and service, both offered and needed?* ACARD's report (1983) identified an important gap between the two for the small firm applying AMT for the first time, with communications shortcomings between supplier and user leading to misunderstandings and equipment problems. Prefer robust, proven technology and machines to new equipment. That also goes for software. It is even more important than for other hardware and software purchases to make as sure as possible that an AMT supplier is likely to stay in business. Get reliability figures and find out maintenance costs for different levels of integration.

Consultants

Japanese companies, replete with engineering skills and fed a diet of eager young engineering graduates, make limited use of consultants in manufacturing. Western firms, with limited resources in-house, often must learn new ways from scratch. Hence the pervasive reliance on consultants, who have never been busier than during the last few years. Consultants have important uses. The best have broad experience, are able to steer an impartial course between rival technological claims and have developed valuable financial and technological assessment skills. They are usually well equipped with business awareness and common sense. But, as in the case of hardware and software suppliers, it is essential to bear in mind what they can and cannot be expected to do. There are many reasons for a company to employ consultants, some of which require the company concerned to swallow the bitter pill of admitting that in some areas it is lacking. A consultancy's greatest asset is its objectivity. It can review an organization's needs, current methods and procedures without fear or favouritism. Its second greatest asset should be depth of knowledge. This is where careful selection of a consultancy becomes important. If a company is engaging a consultant to supplement its own lack of knowledge then it must make sure that the consultancy has a good understanding of not only the technology in question but also the manufacturing processes and the competitive climate within that manufacturing sector. A consultant should be able to advise on best practices within particular industries, without revealing trade secrets. A consultant's third greatest asset is time – the time to devote to developing and, possibly, implementing a particular project, the time that in-house management cannot afford.

Most consultants know what they are capable of achieving but, unfortunately, most clients do not always know what they want, and it is here that the trouble arises – right at the start of the project. One area of Japanese management methods which can translate effectively into a western business culture is the obsessive attention to planning and preparation. Buying in any service or skill should be preceded by a thorough analysis of needs in order to provide specific terms of reference to a consultancy.

Management consultancies are business like any other. The industry has undergone considerable structural change in the last few years. The traditional consultancy providers, some broad-range, others 'boutiques' specializing in areas such as strategy, finance, information technology and manufacturing, have been joined by the large international accountancy firms which have been forced by increasing competition on the audit side (and client resistance to ever growing fees) to look for other sources of business. Like the computer hardware manufacturers, they are moving out of the back office, typically into office systems and, increasingly, into manufacturing automation. At the other end of the scale, university and polytechnic engineering departments have been driven by the tough financial times to enter the consultancy arena. There are no hard and fast rules for calculating which is best. Apart from cost, it is worth checking which consultancies are most active in your area of business, how much of their business is in manufacturing, and the background of the principals. Not all firms have the same degree of experience in manufacturing and IT. Some consultancies have agency or commission arrangements for hardware and software packages, others have in-house analysts and programmers to keep busy. Their advice might not be truly objective as a result.

The best consultants are valuable as change agents, trainers, financial and technological modellers and facilitators in implementing new organization or technology. But the company must manage its advisers, not the other way about. However good, a consultant cannot compensate for lack of commitment on the part of the resident management, nor can he or she turn a hurried, badly planned project into a good one. Managers can delegate authority, but not corporate responsibility to a consultant, who in any case would refuse to take it. At the end of the project the consultant walks away from it; the company must live with it and turn it from a project into a daily part of its operating success. The best consultants recognize this process and focus on achieving the transfer of ownership of the new facilities from consultant to client in the smoothest and most motivating way. They know that most of their work is repeat business from satisfied clients who use consultants to continue to grow profitably.

How to work with consultants

Terms of reference

These should include:

- A short statement setting out the background to the requirement. This will ensure that the management and the consultancy have a clear understanding of the context in which the assignment will be performed;
- The objectives of the assignment, both tangible and intangible;
- Its scope;
- When the job will start and how long it will last;
- The cost of the job and the basis of charging;
- The experience of the consultant who will do the job;
- A timetable of actions, phased to suit the specific business conditions and detailing the resources that will be provided by the consultancy or will be required of the client during each of the phases;
- The willingness of the consultants to assist, if requested, with the implementation of their recommendations;
- Any special terms of business (such as conditions for early termination).

Preparing the ground

- Prepare the ground by telling your managers and other employees why consultants are being employed, when they will arrive, what they will do, the type of information they will require and what are the expected benefits from the assignment. It should be made clear that the consultants have the backing of the management and will require the full cooperation of every member of the organization with whom they need to discuss the project.
- Appoint a senior person within your organization to act as the consultant's main contact, so that there is an established and recognized channel for obtaining information, reporting progress and any difficulties.

continues

continued

- Talk to the consultant who will carry out the assignment and make sure that he or she has achieved a full understanding of the requirements and has all the facilities and information needed to get down to constructive work as soon as possible.

The assignment

- It is essential that the progress of the assignment is closely monitored against the planned timetable;
- Insist on regular verbal or written reports;
- Discuss unexpected difficulties;
- Ensure the consultant is getting access to the right people;
- Pay particular attention to the transition points between phases of the job, especially to ensure that momentum is not lost;
- Check that all necessary decisions and actions are being taken in good time;
- Allow sufficient time for report preparation and discussion;
- Be prepared to replan if it is shown to be necessary.

Follow up

At the completion of the consultant's assignment, it is important to review the results. If any of the final recommendations come as a surprise, then it is probable that the work has not been properly monitored.

Examine carefully the advice offered, discuss the findings with the consultants and ensure that there is a detailed programme for the work remaining.

After the consultants have left, there is always a danger that the benefits will be eroded by neglect or lack of control. To help avoid this, check that new developments and procedures are being properly applied and results reported to you regularly and that they are not being undermined or duplicated by continuing old methods or concepts. It may be helpful to arrange follow-up visits by the consultants to review implementation progress and to discuss any difficulties that are arising.

Source: Computing Services Association Briefing Note.

Building internal resources

All these parts of the scattered manufacturing database – written sources, official programmes, customers, suppliers, equipment makers, professional institutes and consultants – hold valuable clues to what makes modern manufacturing work. But the assistance must be on your terms, not theirs. There is

no ultimate organizational or technological goal which, when reached, will afford permanent competitive advantage. Computer integration itself is just a way station, an advanced way of switching information around which may or may not be cost-effective at a certain stage in permitting new processes, products and organizational forms. Since there is no final destination for the manufacturing odyssey, the emphasis switches from the arrival to the journey, from fixed objectives to trends, from blocks of knowledge to information flows, from control to adaptation, from accomplishing set tasks to challenging their purpose and developing new ones. This behaviour has been christened 'flow-state management' to contrast it with the solid-state management of the past (Ackerman, 1984).

This is the framework within which you can use the helpers for specific projects. Consultants and equipment suppliers can and should be used as catalysts for change, but the main ingredients for success must be in place already. This is why proper planning and cost justification are so important. Beware, therefore, of justifying investment in new equipment just because it qualifies for official funding or a supplier offers discount terms as a proving site. Projects should make commercial sense unsupported. Any extra funds or support can then be used to reduce the risk.

Before putting too much of your strategy in other people's hands, it is worth pondering a final point. The world's best manufacturing nations are probably West Germany and Japan. Both countries use state-of-the-art equipment, and use it extremely well. But it may be no coincidence that their manufacturers make a lot of their own capital equipment and pride themselves on modifying and improving the machines that they buy in.

According to Hayes and Wheelwright (1984), in 1980 60% of Japanese industry's capital investment budget went on upgrading old machines, compared with 25% in the USA. Competition from their customers has the side effect of stimulating the West German and Japanese machine tool industries to ever greater innovation. More important, it leaves manufacturers themselves, rather than their suppliers, in control of factory processes and improvements to them. And even if the processes are themselves standard, the modifications (accumulated learning) built into them may give a proprietary edge which is hard to copy.

The immediate need to apply new manufacturing technology or information systems may be urgent and compelling. But they must be set also in the long-term context. Consultants can make a company aware of its unhealthy lifestyle and prescribe a new regime; equipment makers help the corporate fitness programme to get off to a good start. But neither can take the place of the firm's own commitment to the world-class manufacturing effort. In the end there is no substitute for the steady building of the company's own manufacturing muscle: a highly skilled, flexible and committed workforce; lean, non-hierarchical and innovative management; an ever growing core of product and process know-how; and obsessive attention to customers, competitors and manufacturing targets.

References

Abe K. (1987). (From) How the Japanese see the future in JIT. *Just-in-Time Manufacturing*. Berlin: IFS, Springer-Verlag

ACARD (1983). *New Opportunities in Manufacturing: the Management of Technology*. London: HMSO

Ackerman L. (1984). The flow state: a new view of organisations and managing. In *Transferring Work* (Adams J., ed.). Miles River Press

Gunn T. G. (1987). *Manufacturing for Competitive Advantage*. Ballinger Publishing Co.

5

IT investment

In 1993, lack of investment was seen as the biggest factor preventing UK manufacturing from being world class. This was closely followed by traditional organizational structures and culture, which many managers in manufacturing saw as restraints to growth. Firms also saw themselves as vulnerable on R & D and the use of advanced manufacturing technology relative to Germany's performance which, despite the changes to the German economy since the late 1980s, was still regarded as the biggest competitive force within the EC (*Manufacturing Attitudes Survey,* 1993).

Despite a healthier attitude to long-term planning in the West – most US companies work to a five-year plan but only approximately half of European companies plan that far ahead – there is still a reluctance to regard the buying-in of technology as an investment rather than a cost.

According to several surveys, investments in IT have failed to live up to expectations and one comprehensive piece of research by the Kobler Unit at Imperial College, London (Hochstrasser, 1990) found marked disagreement between managers about the true value of large-scale IT investments. More than a quarter of those interviewed admitted that they were not disciplined enough to monitor continually how the return on capital of their IT investment compared with that of their other investments. Of those that did follow strict procedures, one in eight was positively dismayed with the payback obtained. The study further investigated the reasons why companies invested in IT in the first place and, indeed, why they would continue to invest in something that they, in the main, seemed to regard as a necessary evil. The reasons were varied but, significantly, only 16% invested after fully calculating the benefits (see Figure 5.1). And yet, IT spending is predicted to rise considerably in the

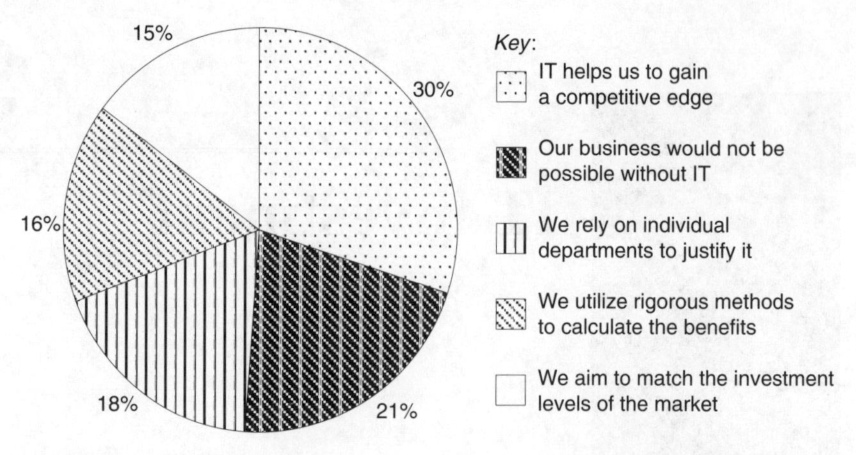

Figure 5.1 Reasons for investment in IT. *Source:* Hochstrasser and Griffiths (1991).

next decade as more advanced technology becomes available and fiercer market forces dictate its implementation.

The *Manufacturing Attitudes Survey* (1993) shows that UK manufacturers believe that they have made real strides in quality and competitiveness.

> 'But it is still worrying that cost is still seen as our competitive edge, rather than excellence in other areas. R & D, for example, is seen as a weak area, with most respondents identifying it as a cost rather than an investment. Limited resources are put into it as a result, so it takes us longer to get there and we make less money. The lack of investment identified by industry as holding it back is a major reason for this cost-obsession. Unless we tackle it now, it will be a major impediment to UK manufacturers becoming world class.'

Value and control

There are two factors which seem to worry manufacturers the most – how to get value for money from IT investment and how to control the cost of IT once they have got it.

Hochstrasser and Griffiths (1991) suggest that many companies do not, in the first instance, appreciate the true costs of installing IT and therefore are not prepared for the total spend involved. In their book they offer a checklist of common cost elements to be used when proposing and evaluating IT investments.

Assessing IT costs

Hardware costs

Mainframes, minis, PCs, display units, primary and secondary data storage, printers or plotters and accessories.

Software costs

Off-the-shelf packages are cheapest but bear in mind there may be some modification costs, and also installation costs.
Application generators. These can be up and running in a very short time if the skills are available in-house. If not, budget for an outside consultant to come in and do the work.
Custom-made software is the most expensive option but may be unavoidable. Requirement have to be carefully planned in order for the eventual performance to match needs. Consultant time again could be expensive.

Installation costs

A huge variable depending upon the application and the hardware/software environment in use.
If a paper-based activity is to be computerized, the costs of inputting current records needs to be taken into account.

Environmental costs

IT installations often necessitate changes in the physical environment such as underfloor wiring, extensive cabling, air conditioning, dehumidifers, new lighting, new factory layouts or additional furniture.

Running costs

Combined power consumption of all new technology could be considerable. This is one area where it is worth having an energy audit to see if savings can be made.
If external databases are accessed, access fees, computer charge times and telephone bills should be included.

Maintenance costs

Service and maintenance support has to be planned for. It is advisable to take out a comprehensive maintenance contract with a supplier as equipment breakdown can prove very expensive to the business.

Security costs

Security procedures need to be built into any operation. A contingency plan needs to be formed to cope with disasters. Duplicates of essential material need to be maintained, preferably at an off-site location.

continues

continued

Networking costs

Local and wide area networks (LAN/WAN) to share both information and hardware devices such as printers or data storage systems often need the additional expenditure of dedicated workstations.

Training costs

Initial retraining of personnel to cope with advanced manufacturing technology could be a large up-front investment. But training should be an ongoing activity so that users develop in pace with the IT developments.

Wider organizational costs

Incompatibility costs: the implications of not buying compatible equipment need thorough and serious consideration.
New salary structures: training people to use – and possibly program – sophisticated computer equipment will result in upgrading their jobs and salaries.
Transitional costs: the introduction of new systems often incurs costs because of temporary job interruptions as prospective users are unable to maintain their normal degree of productivity while learning how to use new equipment.
Management costs: considerable management time may be needed to be spent on meetings, courses, training, security reviews, audit verification, counselling and explaining policies.

Source: Hochstrasser B. and Griffiths, C. (1991).

IT investment appraisal

'There is no such thing as a generic "IT" project – IT is a catalyst that facilitates the achievement of business goals' (Potter and Rosseinsky, 1992). Potter and Rosseinsky of Coopers & Lybrand further argued that there is no role for IT projects, and therefore IT investment appraisal, but just for business projects and business projects appraisal generally. In their paper they outlined a successful approach to investment appraisal by addressing the following points:

- The benefits of a significant proportion of IT investments are long-term and need to be linked to business plans;
- The short-term benefits of a significant proportion of IT investments are apparently non-measurable, that is, 'soft' or 'intangible';
- The benefits of investing in IT infrastructure, that is, the environment that makes other IT projects feasible, are difficult to identify;

- There needs to be a vehicle for communication of the business benefits of IT solutions.

However, Kit Grindley (1991) in his book based entirely on opinions given by the 5000 IT executives who are members of the Price Waterhouse International Computer Opinion Panel, stated that 83% of IT Directors admitted that the cost/benefit analyses supporting proposals to invest in IT are a fiction.

> 'The well intentioned assertion that you can do a cost/benefit analysis of each system, and decide whether to go ahead, does us a disservice. Because, finally they're all integrated. You have to take a holistic view. What is the totality of 25 systems worth to us?'
>
> (Grindley, 1991)

Yet there are many consultants who do extol the benefits of undertaking a formal appraisal. Potter and Rosseinsky (1992) stated that 'selecting the most appropriate investment option is a question of formally quantifying the costs, benefits and risks'. Costs have been discussed earlier in this chapter; benefits, it is postulated, have to be planned by first investing in risk countermeasures (ensuring that no project fails through lack of human, technical and organizational skill and support), careful monitoring (constantly reviewing the processes and enhancing where necessary) and delivering the benefits.

A 1989 study by the British Institute of Management showed that there was a need to integrate IT with the social and political climate within a company because introducing IT into a company always results in substantial additional investment in continuous management education, development and in reacting to social and political change. The three-year Kobler Unit Study (Hochstrasser, 1990) came to the conclusion that introducing IT often forces a company to redesign completely the shape of its organization and that, often, the payback from an IT system depends on successfully mediating between groups of people who have different interests, particularly when changes are proposed in the distribution of corporate power relating to the control of vital information.

Many experts have suggested that this is the true secret of the success of the Far Eastern countries in merging advanced technology into their manufacturing processes – the fact that they have a corporate culture which is already structured to accept the imposition of technology without causing great social and political changes within the organization. And this, alas, is something which does not cross cultural boundaries. One thing western businesses can adopt successfully from Japanese management methods is the discipline constantly to control, monitor and evaluate.

The lack of relevant and regular evaluation procedures leads to loss of control of IT investments. Without relevant evaluation procedures, the introduction of IT is based on an act of faith; without repeating these procedures at regular intervals, benefits once achieved may no longer be realized (Hochstrasser and Griffiths, 1991).

Specific evaluation techniques

Customer resource life-cycle (CRLC)

A method evaluating the worth of an IT system by the level of its worth to the customer. CRLC gives an 11-stage life-cycle that measures the degrees by which a customer's perception of a product changes over a period of time and how, by introducing an IT system to improve quality and service, it will secure customer loyalty at the critical later stages of the life-cycle.

Value chain analysis

A technique which evaluates the potential of IT to create a 'value chain' of increased productivity, better utilization of resources or more effective information gathering and sharing.

Cost benefit analysis

Various well-established techniques such as 'direct labour cost' and 'stages survey' which seek to evaluate the benefits of IT based on increased data output and cost elimination.

IT strategic grid

A diagrammatic approach to determining where IT investment has been made and where it should be applied.

IT investment mapping

Another diagrammatic approach which puts present and future business strategy against current and future IT investment strategy and highlights areas of strength and weakness.

Critical success factors (CSFs)

Willcocks (1992) described this technique thus: 'a CSF might be used to establish key business objectives, decompose these into critical success factors, then establish the IS (information strategy) needs that will drive these CSFs'.

Risk evaluation

CRAMM (CCTA Risk Analysis and Management Method) is approved by government departments as the preferred method for identifying justified security countermeasures for the protection of their own IT systems that process unclassified but sensitive information. It is a time-consuming methodology, not popular with industry, because a full CRAMM review is expensive and requires experience and knowledge to interpret the results.

Multiple methodology

Internal and external analysis of an organization's needs, methods and opportunities by using several types of evaluation techniques, each starting from a different point within the organization, thus, eventually, revealing the gaps and overlaps in the system.

Methods of evaluation

Hochstrasser (1990) attempted to evaluate IT investments by matching specific evaluation techniques to distinct types of IT projects while, at the same time, stressing that 'it is often not possible to regard IT evaluation as being based exclusively on spelling out bottom line benefits in well defined financial terms'.

Traditional accounting methods are labelled by Hochstrasser as 'hard' – other evaluation methods as 'soft'. In other words, instead of matching costs to profits, 'soft' evaluation procedures match the costs to well-defined management goals.

Hochstrasser defines projects as follows:

- Infrastructure – hardware or software systems installed to enable the subsequent development of front end systems;
- Cost replacement – IT systems introduced to automate manual activities;
- Economy of scale – IT systems introduced to allow a company to handle an increased volume of data;
- Economy of scope – IT systems introduced to allow a company to perform an extended range of tasks;
- Customer support – IT systems introduced to offer better services to customers;
- Quality support – IT systems introduced to increase the quantity of the finished product;
- Information sharing and manipulative – IT systems introduced to offer better information sharing and information manipulation;
- New technology – IT systems introduced to do things that were not possible before and allow a company to exploit the business potential of such innovation.

Some of these projects, for example cost replacement, can be evaluated using the traditional cost–benefit analyses techniques, that is, calculating the money saved in replacing direct labour with IT.

Others can be measured by the amount of extra business generated without adding to a company's resources. Hochstrasser cites electronic data interchange (EDI), point of sale (POS) and just in time (JIT) as examples of economy of scale projects which increase a company's ability to speed up the business cycle, and thus turnover, while using the same level of resources.

Some projects require strong market analyses information to aid the evaluation process. IT systems introduced to extend a company's range of services or products require detailed knowledge of the potential market impact of such innovations before an evaluation of the payback can be assessed.

The use of the customer resource life-cycle (CRLC) technique is suggested as a method of evaluating customer support projects. CRLC is based upon research which showed that it is five times more expensive to gain new customers than it is to keep existing ones. But existing customers' perceptions of a product undergo several changes during the life of a product, and these can be plotted on a life-cycle. Introducing new services/products at a critical stage of that life-cycle can consolidate a market share and, therefore, payback on the IT can be measured in terms of continuing customer loyalty *versus* the cost of courting new customers.

Traditional value-added methodologies can be used to determine the value of IT systems introduced to increase the quality of a finished product (see Figure 5.2). Value linking, for example, is a technique which identifies the impact of a proposed investment by showing, in graphical terms, the links between IT and the areas of the business in which it produces value.

It is new technology projects that are exceptionally hard to evaluate containing, as they do, a high element of risk. Hochstrasser suggested that risk minimization strategies and risk evaluation techniques be deployed, which must take into account the insecurity of the investment while balancing it against the potential rewards generated by offering a strategic advantage over competitors and a greatly increased market share.

Build in communication

Coopers & Lybrand have argued that there needs to be a vehicle for communication of the business benefits of IT solutions and that much current practice in justifying IT investments fails to address these points and thus often leads to inappropriate decisions being made.

This needs to be addressed in two ways. First, IT investment needs to be linked to a business strategy with fully defined goals set along the way which need to be met. Second, once that has been established, a mechanism needs to be built into the process whereby success in meeting goals can be measured and reported.

Investment appraisal, in Coopers & Lybrand's experience, is a process that consists of the following:

- Understanding how proposed options for investment address business needs;
- Making a decision on which options to adopt;
- Establishing a framework of responsibility and accountability to evaluate the achieved benefits and costs of investment;
- Using the evaluation framework to identify actions that generate business benefit;
- Monitoring and managing the achievement of those benefits.

The process can be logically broken down into four elements:

- The business case – to structure the problem of option selection;

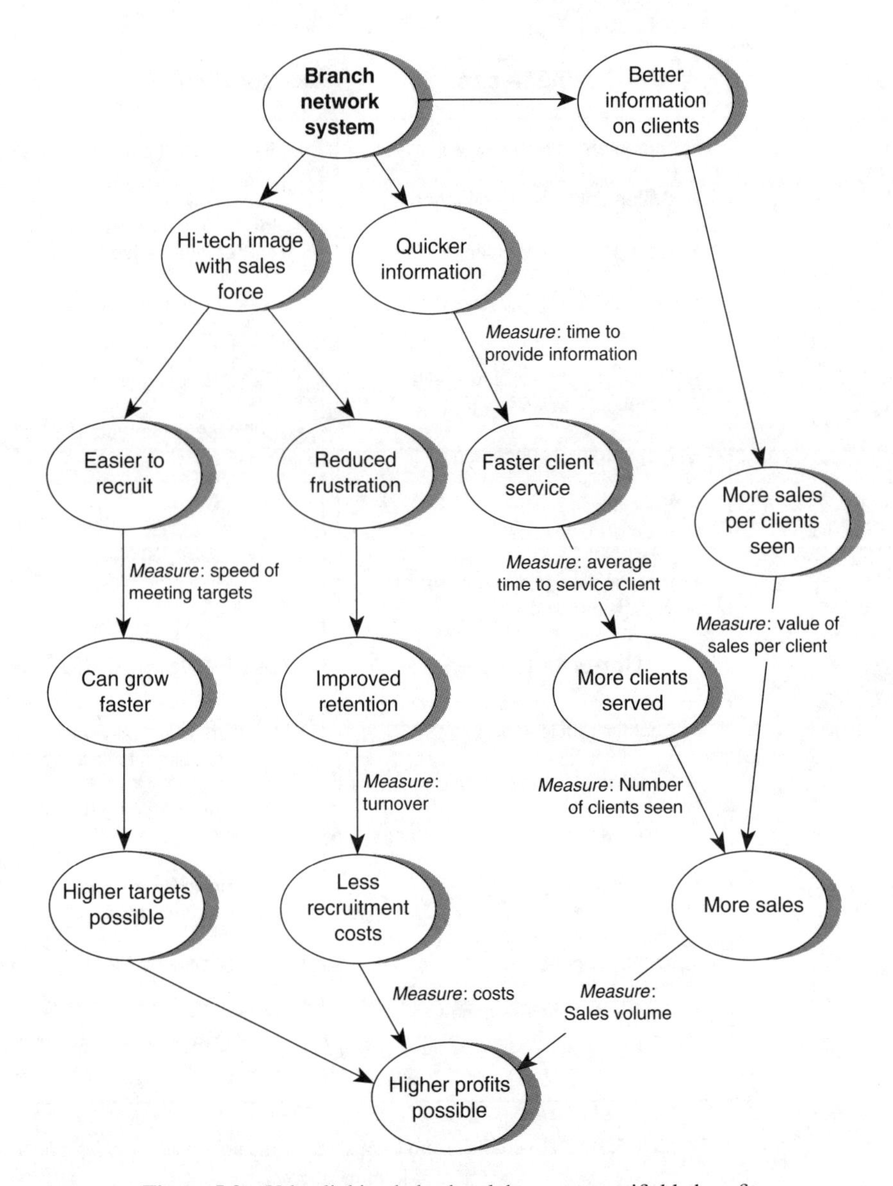

Figure 5.2 Value linking helps breakdown unquantifiable benefits.
Source: Coopers & Lybrand (1991).

- Cost–benefit–risk analysis – to decide on the best option;
- Benefits planning – to establish responsibility for benefits;
- Post-implementation review – to ensure that benefits are obtained.

Justifying investment is easier if claimed benefits can be quantified in the common language of finance.

Business case	Cost/benefit-risk analysis
• Statement of business objectives • Statement of likely scenarios • Options to be considered • Benefits, costs and risks for each option • Match of option features with benefits • Major areas of uncertainty • Specification risk counter-measures • Identification of appropriate appraisal method	• Evaluation of benefits, costs and risks • Identification of the value of further information • Risk counter-measures for technical delivery • Benefits documentation • Full case documentation
Benefits planning	**Post-implementation review**
• Benefits profile (when) • Benefits allocation (to whom) • Benefits responsibility (by whom) • Benefits measures • Risk counter-measures for benefit delivery • Review of business case	• Measure of actual versus expected benefits and costs • Responsibility for corrective actions • Responsibility of opportunities for increased/decreased investment • Review of need for renewal/new solution • Identification of ways to improve future appraisals

Figure 5.3 Deliverables from each stage of the appraisal. *Source:* Coopers & Lybrand (1992).

Harrison (1990) echoes these sentiments in his book:

'Financial and strategic appraisal are inevitably interlinked, the latter providing a context and information for the former.... Financial appraisal must not be seen as an alternative to the strategic appraisal of an investment.'

But Harrison, nevertheless, devotes a great deal of space to conventional investment appraisal methods such as payback period and discounted cash flow – appraisals which have been constantly discredited by many other writers and researchers.

Hill (1985) quotes authors from as far back as 1974 who 'illustrate the simplistic nature of the accounting perspective and challenge its unevaluated application'. (See also Chapter 10 of this book.) He outlines in one chapter the difference in western and Japanese companies' approach to investment by quoting from K. Ohmae's paper 'Japan: from stereotypes to specifics'. Ohmae discusses investments made by various Japanese companies and poses several questions.

> 'How many contemporary US corporations, relying on Return on Investment [ROI] yardsticks would have embarked upon the development of a business that required a twenty year incubation period, as did Nippon Electric Car Company with its computer and semiconductor business?... Would Honda have so obstinately persisted in using its motorbike profits to bring its clean-engine vehicle to market if it were a corporation that measured the ROI of each product line and made its decisions accordingly? In fact would any manufacturers be entering the four wheel vehicle market in today's environment if ROI were the investment criterion?'

However, while Hill makes a strong case for abandoning traditional accounting methods, he also makes an equally strong case for not ignoring intangible, difficult-to-evaluate reductions or benefits which accrue from investments. To do so would be a dangerous mistake and, he further adds, to combine that with the mistake of underestimating the working capital needs and infrastructure costs associated with each proposal is a recipe for disaster.

The key to effective decision-making with regard to IT investment is to create an environment that generates information which is of value to the strategy-builder. Hill defines this information as business- and performance-related financial information, which a company can then use as tools in the formulation of a business and investment strategy. In other words, an ideal partnership between accounting and manufacturing.

Not just evaluation but a strategy

A 1988 paper by Glen Peters was based on the results of a review of over fifty IT projects in some of the world's leading companies and Peters used his findings to illustrate the concept of investment mapping.

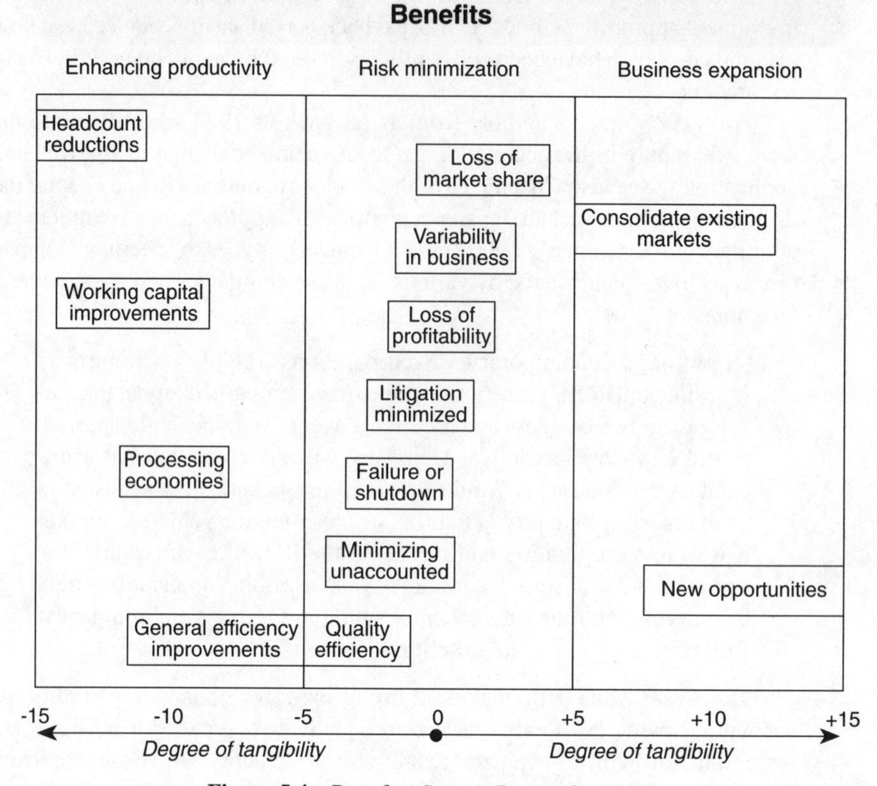

Figure 5.4 Benefits. *Source:* Peters, G. (1988).

Peters found that he could divide the expected benefits of an IT investment into three categories: enhancing productivity, risk minimization and business expansion. As one can see from Figure 5.4, the benefits are described in business not financial terms. Peters argued that once the business benefit has been carefully defined, it is then possible to calculate 'payback'. The degree of tangibility shown at the foot of the table does not denote negative or positive benefits but, rather, the degree to which a 'hard' benefit or 'payback' can be identified.

One may not agree with Peter's assessment, of course. Most manufacturers would be able to calculate quite easily the amount of money that their companies would lose per day in the event of a failure or shutdown – yet Peters places it at around –2 degrees of tangibility. Five years on from the Peters study, IT manufacturers have and are investing heavily in a perceived need for fault-free products, thus demonstrating the increased customer demand to invest in risk-minimization IT.

Nevertheless, the Peters study went further: 'Benefits were not the only criterion used to appraise the value of an IT project,' he stated, 'The relation-

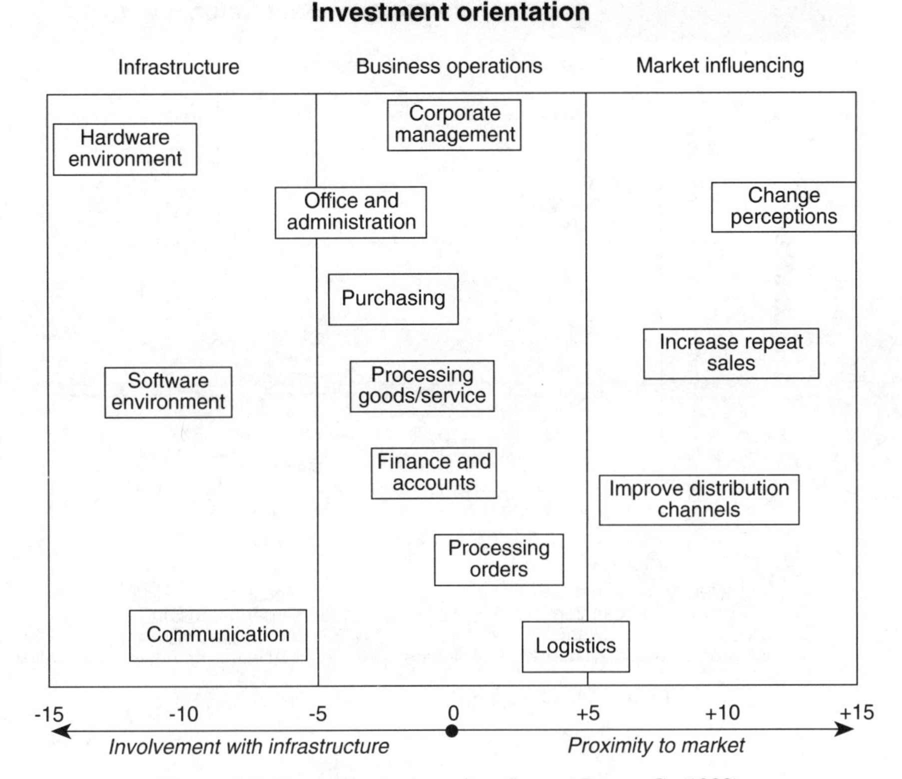

Figure 5.5 Investment orientation. *Source:* Peters, G. (1988).

ship or orientation of the investment towards the business was also frequently used in evaluation.'

Again, he divided this criterion into three categories: infrastructure, business operations and market influencing. He noted that, among the companies interviewed, approval of expenditure on infrastructure projects was the least popular and therefore any such need was more likely to be tacked onto a market-influencing project to get it past the board.

Figure 5.5 shows the Peter's evaluation of investment orientation. Using this system, it is therefore possible to draw up an investment map which strongly reflects a company's business strategy. It is a useful way of comparing one proposed investment with another and assessing the possible benefits and the orientation to the business. A sample investment map is shown in Figure 5.6.

As with all methodologies, investment mapping is an aid to decision making, not a blueprint which cannot be altered, or a rigid formula for calculating levels of investment.

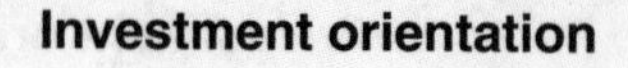

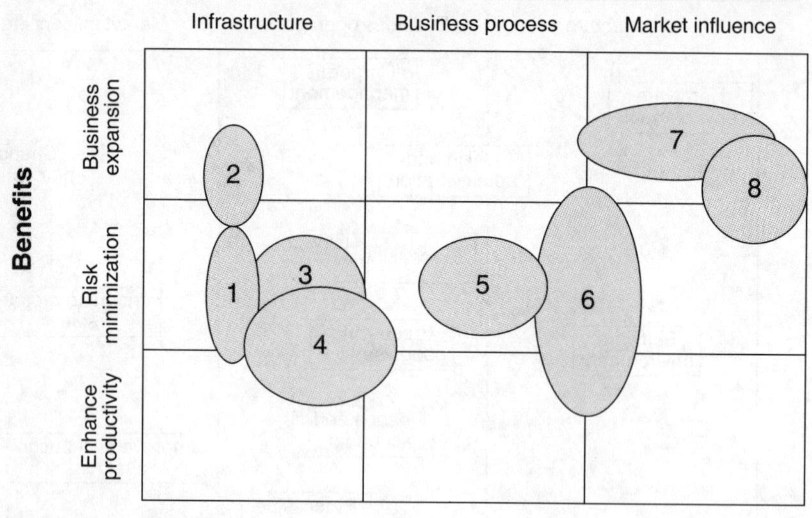

Key:
1. Migration to database environment
2. ATM network expansion
3. Office systems network
4. Electronic data interchange for suppliers
5. Accounting
6. Order processing
7. Automated insurance
8. VANs for the business traveller

Figure 5.6 Investment map. *Source:* Peters, G. (1988)

Cost escalation

The 1990 Price Waterhouse Survey, which formed the basis for the Grindley book (1991), found that cost escalation of IT spending was a serious concern for most of the respondents.

The four main causes appeared to be:

- The difficulty of estimating and controlling software development costs;
- The failure to appreciate how much hardware is required;
- Coming to grips with users' requirements as the system is being built at the time work starts, instead of before;
- Hardware and software becoming technically obsolete every three to four years.

The Kobler Unit Report (Hochstrasser, 1990), on the other hand, found that the largest uncontrolled IT costs stemmed from:

systems and IT departments, like rarified R & D units, are now required to tak
a holistic approach to technology and embrace the business strategies tha
govern everyone else within an organization.

This evolution has, of course, been given a name – hybridization. The
description refers to the happy marriage of business and technical expertise tha
the modern hybrid manager should possess.

A review of IT expenditure within the UK conducted by *Computer Weekly*
(1993) estimated that £15,897 million was spent in 1993 and this will increase
sharply to around £18,000 million in 1994. Yet, conversely, research from
London's City University (1992) suggests that over a fifth of this outlay is
wasted and between 30% and 40% of IT projects produce no discernible
benefits at all.

Another survey by Business Intelligence (1992) showed that organizations
that have employed hybrid managers or retrained existing personnel have
achieved remarkable results in controlling budgets and schedules. Esso UK, for
example, found that the number of projects completed on time rose to 90%
after a hybrid training programme was introduced. The survey quotes many
international organizations as believing that hybrid managers can be a potent
force in helping companies make profitable use of IT.

In the USA, it has long been recognized that IT managers must be 'team
players' and can no longer function in a vacuum of technical excellence, unable
to relate to other board members or to cultural and organizational changes.

Fortunately, most centres of learning now recognize the need to train
hybrid managers from the beginning and combine business and administration
skills with computing courses. One such programme, devised by Lucas
Industries and other companies in conjunction with Warwick University,
covers topics such as business strategy, marketing and influencing skills,
through to decision support systems, software technology and systems
analysis. In the words of Bob Galliers, the course professor at Warwick:

> 'Perhaps in IT we've had our sights set too low in terms of our ability
> to make a major impact on an organization. But what we are now talking about is the development of people at the highest level of business
> decision making with an understanding of what information technology can and can't do.'

The benefits of developing/employing hybrid managers are defined by
Business Intelligence (1992) as:

- Strategic (new IT opportunities identified; systems more in line with business requirements);
- Operational (more projects within budget and to schedule; higher project success rate);
- Cost effective (development costs recovered more quickly; falling corporate computing costs);
- Personnel enhancing (improved job satisfaction within IT departments; fewer computing staff recruiting problems).

- Rising salaries of IT support staff;
- Man hours spend on the installation of new systems;
- Man hours spent on the configuring or reconfiguring of software;
- Money spent on a new office/factory environment to accommodate IT;
- Add-on hardware such as PCs, printers and other accessories;
- Loss of business due to down-time;
- Man hours spent on learning or evaluating systems;
- Consultancy and training fees.

Other writers also agree that investing in IT is useless if the systems and business functions are separate and that an organization should adopt business management, rather than technical elegance, as the driving philosophy of IT. It is easy to blame technological inadequacy for corporate underperformance. Nine times out of ten it is not technology's fault. The bulk of poor exploitation of IT is caused by poor business management.

Then again, the technologically backward manager, who knows nothing about the capabilities of technology in relation to his or her company's core

Case study: Black & Decker

Black & Decker started using electronic data interchange (EDI) in 1987 in response to pressure from its customers – the major DIY chains in the UK. The company was also installing JIT manufacturing techniques and EDI was complementary to that. It transformed the salesperson's role overnight, making him or her a true salesperson not just an order taker.

The company has large manufacturing plants in Germany and Italy as well, and the UK and German plants have followed parallel paths, developing similar and highly complementary EDI systems, though managing them separately. Every company that supplies to Black & Decker UK or Germany has been put on to EDI.

The only difficulty has been with standards. Black & Decker have had to link up to INS and IBM, Tradacoms standard and AT&T Istel Edifact standard, in order to accommodate the systems of suppliers and customers in the UK and abroad. 'We see the Edifact as the standard for international work. Most countries are moving towards it and it will be the standard in five to ten years time,' said Black & Decker's European IS/EDI manager, Ken Elleson.

The company claims that EDI has not entirely justified itself in terms of cost benefits, because of the high cost of entry into the networks and the cost of transactions, as well as the patchy adoption of EDI among Black & Decker's customers and suppliers.

'There have been intangible benefits,' admitted Elleson. 'The value of EDI is that it is a better way of doing business, rather than a low cost method.'

activities, is the one most prone to 'IT fever' – the desire to purchase more and more technology because he or she is easily persuaded that it is indispensable. There are two types of IT fever: one is the fetish for everything new and the other is the creeping paranoia that everyone else has it except the sufferer. Both types of the malady, joking aside, are dangerous to a company's economic welfare since, like the alcoholic who refuses to think about why he or she is drinking, the IT fever victim refuses to evaluate before spending.

The technologically backward manager is also prey to demanding IT wizards within the company, who push for more and more gadgets to play with irrespective of the cost-effectiveness of more advanced technology within the business, and to the persuasiveness of IT sales staff and consultants.

The case for an IT director

In 1992 it was estimated that approximately 60% of IT directors in the UK had a seat on their company's board. This is a dramatic leap from ten years ago when fewer than 20% were in such an exalted position. But it is still less than many other countries, including the US, Japan and Germany.

However, according to a survey in 1992 by the National Computing Centre (NCC) (Hamilton, 1992), there is a growing trend of IT managers needing to justify their existence within an organization.

> 'Organizations are looking to halve their computing costs. IT managers should be astute politicians and activate those cost savings themselves.
>
> 'The Boards of companies are on a relentless road to cost-cutting. They're not interested in big systems projects anymore. If they can get a facilities management person in to do it at a greatly reduced cost, then that's what they will do.'
>
> (NCC, 1992)

The NCC survey further showed that many companies are breaking up their IT budgets and distributing the funds to different departments, or changing the IT department into a profit centre in order to justify its existence.

Another 1992 survey by Sequent Computer Systems stated that most directors believed that IT did not offer value for money nor did they understand its benefits. Not surprisingly, the survey also showed that board directors and IT managers were seemingly locked in perpetual conflict.

A survey carried out in 1991 by the Oxford Institute of Information Management discovered that successful chief executive/IT manager relationships were based upon a mutual regard for IT as a strategic tool.

The current hot job description in IT circles is 'hybrid manager' – one who has a foot in both the business and IT camps. It appears that this is essential in

the 1990s in order to survive. 'There is no way
performance on the success of the technology,' s
'but if we run out of cash I will be blamed.'

Accountability is the key word. Each econo
resolve to make every part of a company accou
western companies IT seemed to have escaped t
integrate it properly with overall business strateg

The role of the IT manager

The management of IT has evolved into somethi
core businesses it serves, rather than remaining
some elevated technical plane. IT managers, on

The productive partnershi

The ideal IT manager has:

- A background which emphasizes business a within a professional IT career;
- A mission to promote the concept of IT as a mation;
- An ability to contribute to business thinking
- A team role profile which stresses leadersh implementation;
- A very clear understanding of the chief business and about IT;
- An approach to the management of IT w business and of IT thinking.

An ideal chief executive has:

- A career background in general manageme
- A track record of leading organizations thro
- A willingness to be educated to IT, inclu seminars;
- Some personal experience of significant be
- A perception that IT is truly important to mention in the annual report;
- A perception that IT's role is an agent of bu

Source: Oxford Institute of Information Manag

Are IT departments disappearing?

It would seem so, according to recent trends, and IT managers are thus having to develop new roles for themselves – hence the move towards becoming well-rounded executives who can contribute to a company's overall strategy. John Ockenden, the chairman of the UK's National Computer Centre, stated in 1992:

> 'The role of IT directors is being thwarted, first by the technology, such as networking, which is putting the information in the hands of the user. Second, the boards of companies are on a relentless road to cost-cutting. They're not interested in big systems projects any more. If they can get a facilities management person to do it at a greatly reduced cost, then that's what they will do.'

An NCC survey revealed that more and more companies are breaking up their IT budgets and distributing the funds to different business units (Hamilton, 1992).

A *Computer Weekly*/PA Consulting Debate held in London in 1993 among top IT managers showed that intelligent managers were all aware of their vulnerability in the future (Vowler, 1993). 'The survival of the IT department is determined by the importance of IT to the particular business', said one manager, while another stated his confidence in the fact the 'the IT department won't disappear completely', even if the trend towards outsourcing grew. The general feeling was that without a core IT department a company would be at the mercy of external suppliers, so therefore the role of the department and its manager was 'architectural' and to 'become intelligent customers of outsourcing companies ... outsourcing requires a lot of high value management and IT professionals need to grow into that new role'. The consensus was, however, that the leading role that IT departments used to play may have been lost forever. The emphasis now is to break down the barriers around IT and create multifunctional teams. 'Organizations will be more fluid.'

Many state that the radical change has been brought about by a change in emphasis. Business strategy is no longer IT led; it has been down that path with, often, disastrous results. IT is now business strategy led. The IT manager now look for solutions to the core problems of the business and adapts the IT accordingly.

Buying IT in

A survey carried out in 1992 by the Computer Services Association found that 99% of chief executives felt that IT was very important to their companies' operational success and 87% felt it was important to strategic success and that, despite the recession, more companies were expecting their IT expenditure to increase rather than decrease. There was a mixed response to the anticipated use of bought-in services. Consultancy and professional services showed a decrease of 13%, but software development, systems integration, facilities and network management showed an expected increase. The respondents were concerned about the cost of buying in outside services but were appreciative of the flexibility and expertise that they provided. Some expressed concern about loss of control, lack of ongoing support and lack of business knowledge among some service providers.

The big debate in industry in the 1990s has been whether or not to go down the facilities management (FM) route, in other words, hand over the management and operation of part, or all, of an organization's IT to an external source. Organizations have switched emphasis to concentrate on their core activities in order to increase market penetration and become more competitive. The drive for outsourcing non-core activities has followed on from the devolution of management responsibility to departmental, divisional or subsidiary level (Computer Services Association, 1991, 1992).

The current state of the FM market seems to be healthy mainly due to the extremely flexible packages offered to clients. FM suppliers can cover IT activities (computer operation, operational software responsibility, applications software provision and support, network planning, provision and support, hardware provision and maintenance, and consultancy and IT strategy), deal with employment or secondment of client staff, and locations of staff and equipment either on the premises of the client or supplier.

Client companies may retain ownership of their hardware or sell it to the FM supplier, who can also take ownership and responsibility for software, licensing and support and responsibility for the networks.

But the ability to outsource effectively comes back again to being able to determine a company's true cost in-house.

When should facilities management be considered?

- To enable management to concentrate on running core business activities;
- To maintain continual service during a period of change such as:
  - moving from one computing environment to another,
  - changing from central to distributed computing,
  - relocating the computer centre,
  - acquisition or divestment of a business,
  - the creation of a new organization,
  - a major change in trading;
- To account for clearly and control IT costs, and potentially reduce them;
- To provide a cash injection and subsequently pay for the service on a monthly basis;

continues

continued

- To accommodate major technological advances without heavy capital investment;
- To cater for periods of uncertain computer demand;
- To define and implement service level agreements and clearly identify responsibilities and accountability;
- To improve the quality of the service for end users;
- To take advantage of application packages which may run on alternative hardware;
- To avoid the management problems of staff recruitment, motivation and retention;
- To have cost effective access to specialist skills as and when required without the overhead of employing permanent staff.

References

Business Intelligence (1992). Business Benefits and IT – the hybrid skills connection.

Computer Services Association (1991). *An Introduction to Facilities Management*

Computer Services Association (1992). *In-house Computing or Facilities Management – the real cost*

Grindley K. (1991). *Managing IT at Board Level.* London: Price Waterhouse/Pitman

Hamilton S. (1992). Board Games. *Computing,* 10 September

Harrison M. (1990). *Advanced Manufacturing Technology Management.* London: Pitman

Hill T. (1985). *Manufacturing Strategy: the strategic management of the manufacturing function.* Macmillan

Hochstrasser B. (1990). Evaluating IT Investments – matching techniques to projects, Kobler Unit. *Journal of Information Technology*, **5**, 215–221

Hochstrasser B. and Griffiths C. (1991). *Controlling IT Investments.* London: Chapman and Hall

KMPG/Impact Research (1990). Oxford Institute of Management

Manufacturing Attitudes Survey (1993). Computerisation/Benchmark

Ohmae K. (1982). Japan: from stereotypes to specifics. *The McKinsey Quarterly.* Spring, 2–33

Peters G. (1988). Evaluating your computer investment strategy. *Journal of Information Technology*, **3**, 178–188

Potter and Rosseinsky (1992). *IT Partnership Bulletin No. 1 – Investment Appraisal.* Coopers & Lybrand

'Spend shift' analysis of major IT budget holders. *Computer Weekly,* Kew Associates, Autumn 1992

UK IT market analysis. *Computer Weekly* Publications, Autumn 1992

Vowler, J. (1993). Future of IT – where to from here? *Computer Weekly*, 14 January

Willcocks L. (1992). IT evaluation – from price to value. London: City University Business School Working Paper

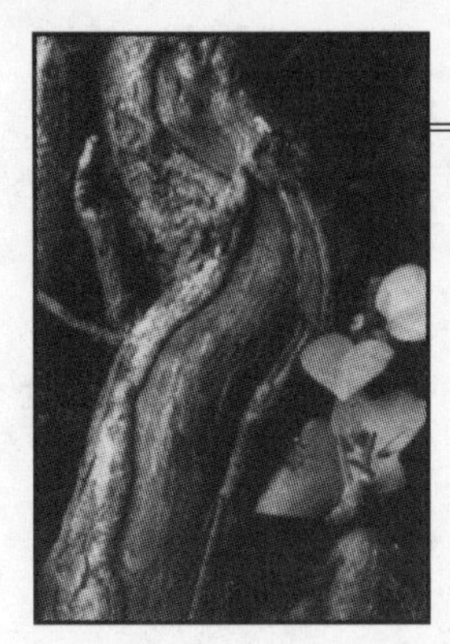

6

Limiting the risk

The nature of risk

This book has put forward several reasons for making simplicity the watchword in two stages of manufacturing systems improvement: first in stripping non-value adding activities out of the process, and second in using appropriate information technology to automate prudently. There is another powerful incentive to keep things simple in the need to limit risk.

There are risks attaching to any physical installations and administrative procedures, just as there are to any property. They include accidental damage through flood, fire and other hazards; deliberate damage such as sabotage or blackmail; fraud; and damage through industrial disputes and strikes. All these are well known to company secretaries, insurers and industrial relations managers, and are taken into account in normal business planning. There are other risks in negotiating change. The greater the alteration of established processes, products and habits of working, the greater the threat of losing customers, upsetting employees and by either of those means disconcerting the bank and shareholders as well. This too is not new, although the increasing speed and incidence of change has made it a factor of growing importance. But the relatively straightforward methods of traditional investment appraisal, insurance and risk limitation are decisively altered by integration and information technology.

Threats to a company's information resource are strikingly omitted from most discussions of information systems needed to support the more ambitious programmes of automation and CIM. Implying that there is no alternative, the 'automate or liquidate' argument scarcely allows for caution. Yet there is a large and growing literature on computer-related crime, and specialist consultancies make a handsome living out of adding security elements to existing computer systems. All the evidence is that information security in its broadest sense is a major issue for most companies even though they do not know it, and that not surprisingly, in the overwhelming majority of cases, levels of security are low, perhaps dangerously so. A report prepared for the European Commission in 1988 found that a sample of 20 large European companies over a cross-section of industries were more dependent on their computer networks than they realized, that their networks were extensively exposed to disruptive events, that only one out of the 20 had anything like adequate security and that satisfactory standards were difficult to achieve. 'We believe', the authors concluded, 'that action is essential, urgent and must be directed at getting the fundamentals right.... A lot has and is being done to equip Europe for the "IT age". But there is no doubt that the study shows that at present the "IT age" is

Network security

The report for the European Commission on *The Security of Network Systems* by Coopers & Lybrand (1988) brought into sharp focus the inadequate protection given to computer information in most companies. The report, costing £1 million and described as the most comprehensive ever, carried out 20 case studies among top European companies over industries ranging from motor manufacturing to banks. Of the 20, only one rated a satisfactory mark across all security aspects. Most companies were found inadequate under four out of eleven assessment heads: administration of security, telecoms, contingency planning and microcomputers; borderline on five others: configuration management, network operations, documentation security, support services and data security; and adequate only on physical security and computer operations. The case studies involved large firms with resources and established systems. They are likely to have better security than less well endowed firms. Yet the safeguards were clearly unsatisfactory. Why?

One of the main findings of the report was that satisfactory security for IT networks is inherently difficult to achieve. The authors identified nine reasons.

(1) *Lack of security awareness.* The 'single greatest complicating factor' is the lack of common understanding of security implications and responsibilities among the people using and making decisions about networks, ranging from general managers to DP professionals to staff operating terminals. 'This understanding simply does not, as yet, exist in practice outside the defence/national security community and one or two notable exceptions elsewhere.'

continues

continued

(2) *Organizational structures.* Networks are no respecters of functional boundaries, and responsibility for their security is elusive. 'Our study shows levels of security are highly correlated with the existence or otherwise of a security manager. To make a broad generalization, no security manager means poor security.'

(3) *Costs.* Spending on security is often viewed as discretionary, and not approved.

(4) *External dependencies.* Few firms have information networks that are insulated from the rest of the world. But the outside world imports massive complication. For example, any network system using public telephone links is in principle accessible by any telephone subscriber (there are 700 million) in the world. Without more decisive action on security, the move towards network standards such as OSI actually increases this kind of vulnerability. Bought-in software applications likewise introduce an element outside the company's own control.

(5) *External regulatory and legislative influences.* Export restrictions, regulations governing interconnects with public networks and types and volumes of information which can be transmitted over international communication links all make security solutions more complex.

(6) *Technical complexity.* Networks systems are both inherently complex and subject to constant change. This makes it impossible ever to test systems thoroughly, and directly increases the chances of disruption through unforeseeable circumstances or combinations of circumstances.

(7) *Dynamic technology.* Fast changing technology means that companies are constantly revising their technological base; technologies are unproved or out of date.

(8) *Lack of standards.* Lack of standards in the areas of international communications, password mechanisms and advanced alternatives to passwords cause problems and high customization costs for today's multinational, multivendor, multiapplication networks. (But see also *External dependencies* above.)

(9) *Lack of tools.* 'There are few security solutions that can be applied universally to a full range of a vendor's products, far less across the sort of multivendor, multicountry environment that is now commonplace.'

Apart from these complicating factors, companies surveyed experienced difficulties in other areas, notably in risk analysis, configuration and change management, contingency planning, distributed software and in dependency on a few key staff.

being constructed on foundations which simply will not bear the load.' A survey carried out four years later by the UK Department of Trade and Industry, which subsequently gave rise to a report on the matter of security (DTI, 1992), found that the risk had grown, not diminished. The increase in risk stemmed from two factors:

- A growing threat due to an increasingly competitive international environment (commercial and industrial espionage);
- A growing vulnerability due to a massive increase in the use of IT in general and networks in particular (increased penetration and also crucial dependence on IT which makes downtime catastrophic).

The growing vulnerability has been magnified by the advent of high cost, low performance personal computers. These, say the DTI report, have magnified enormously the problems of IT security within an organization, both because they are ubiquitous and because they lack basic security controls. Also, the introduction of structured wiring, giving common access throughout buildings, together with the communications bridges (linking networks in different offices together) has led towards enterprise-wide access to IT systems. This connectivity is further increased by electronic data interchange (EDI) connecting systems from different organizations, providing greatly expanded access to IT systems.

Risk can never be entirely eliminated, being inherent in the nature of business. That in itself should not be a major concern. But it can be substantially diminished by intelligent planning of the new systems and by preventive maintenance before the most obvious problems are allowed to occur. This chapter therefore addresses the broad issue of information security in manufacturing under three main headings:

- Systems design
- Implementation
- Operations.

Systems design

The failure of companies and computer manufacturers to address security issues until after the event is another conspicuous example of technology-led rather than information-led development. Few managers consciously 'manage' information or systematically define its value to the firm. Still fewer consider how much of it and what quality is necessary for the functioning of the business, although even without this first step they are prepared to spend large amounts of money on computers and telecoms equipment to handle 'DP needs'. This is putting the cart before the horse. The risks to business information can only be contained and managed within the framework of an overall information strategy. The three components of information security – protecting confidentiality, maintaining integrity and making it available in the right quality and quantity to the right people at the right time – are no more separable from the design of the network than quality from the manufacturing process. Like quality, security is not an ingredient that can easily be inspected in afterwards. It must be inherent in an overall design that is simple, robust and appropriate to its purpose, with due account taken of the issues raised by integration and IT complexity.

Integration

When the trade unions struck at Ford UK in early 1988, prophecies of a long strike conducted in a blaze of media publicity were conspicuously unfulfilled. There was none of the ritual posturing characteristic of British industrial disputes of the 1970s, and the strike was settled in a matter of days. There were good reasons for this. In the effort to meet Far Eastern cost competition, Ford of Europe made determined strides towards integrating its 22 European plants, slashing parts inventories and taking up the slack as it moved steadily onwards with just-in-time manufacturing. The attempts paid off, in the sense that the cost difference with Japanese rivals is now down to $200–300 per car. But for every gain through integration there is an increased risk. A fully integrated system is only as strong as its weakest link. Consider the old unintegrated Ford of Europe. The organization contained so much slack that it could support a strike of weeks in one country without suffering significant effect in the others. The product line was not integrated, differing from country to country. Where factories were producing for each other, piles of inventory insulated them for the rest of the world. This 'insurance policy' of excessive inventory actually encouraged both sides to expect lengthy hostilities before any unpleasant economic effects were felt. Neither side had an incentive to make concessions until the strike started biting.

That kind of luxury – just-in-case manufacturing – is no longer economically defensible. Hence the moves towards increasing interdependence and integration. One of the important features of an integrated manufacturing system is that it instantly highlights problems. This is as true of a multiplant system as within the factory itself. In a completely integrated plant or chain of plants, short of complete duplication of every process, any fault in any part of the system can make it economically necessary to bring the whole operation to an instant halt. Factories are open systems, subject to all kinds of influences external to the firm: changing markets or technology, government regulations, the financial fortunes of suppliers and customers. Internal events can have a ripple effect throughout the supply chain, and as integration increases, so does the potential impact of a change to a single key element.

Planned preventive maintenance and condition monitoring

As part of a risk-limiting strategy, planned preventive maintenance (PPM) in conjunction with condition monitoring (CM) is the way to completely avoid expensive breakdowns in mid-production, which can lose valuable output time and waste expensive raw materials.

PPM is a system whereby parts are replaced automatically to a set schedule which is determined by using CM instrumentation to predict the fault rates of machinery. Manual CM units can be linked up to most machines, although most advanced machine tools have CM built in. The disadvantage of CM equipment is that the output signals are sometimes in analogue form, unreadable to computerized analysis equipment. Even those CM units which produce digital signals are not always easily interpreted. Some companies have been working on systems which integrate CM with the duty cycle of the machine being measured, to produce information which maintenance or production staff can read and act upon.

The main value of a CM and PPM programme is that CM will help to predict the part of the machine that is most likely to need replacing and when that point will come and the PPM back-up will then ensure that part is in stock and can be replaced at a time when shutdown is least inconvenient. The correct investment in CM and formulation of a PPM programme will help to eliminate down-time, eliminate overspending on unnecessary parts and overstocking of those parts.

There is a trade-off between systems resilience and integration. The issue is especially important for computer hardware and software. One of the major goals of integration is to increase flexibility and speed of response, and well designed computer software can be an important contributor to these ends. But trouble can arise if the software which gives flexibility for a certain set of conditions is integrated in such a way that it compromises the long-term flexibility of the programme as a whole. This can happen if the software is too integrated, making it impossible to alter the smallest part of the system in isolation from the whole.

> 'Thus, when a new set of requirement evolves, caused perhaps by fundamental shifts in markets and product lines, the organisation might find that it is unable to respond quickly because the integrated software systems do not lend themselves to change. The potential need to "unintegrate" and redesign part of the CIM program was not recognised as an area of risk.'
>
> (Kimmerly, 1986)

Faced with a choice, the technologist's natural tendency is to go for maximum integration. For the specialist, working with the state of the art is an end in itself. The proper business response should be extreme scepticism. The greater the degree of computer integration, the greater the difficulty of actually achieving all the theoretical benefits and the greater the risk of real failure. Whatever the benefits, CIM increases the risks. Secure, resilient solutions manage information communications on a need to know basis, with very limited integration. Consider the cell organization of the IRA (Irish Republican Army), a resilient operation if ever there was one. At the present stage of development, for most businesses it is almost certainly preferable to sacrifice some of the theoretical benefits of complete hardware and software integration for a simpler, more modular approach in order to make the system as a whole more fault tolerant and robust.

Computer complexity

IT is likely to be used at some stage as part of the integration process. But over and above their role as integrator, computers, if allowed to (as they usually are), drive a remorseless cycle of increasing complexity. As J. A. Schweitzer (1986) put it: 'The quality and consequently the value of business information increases exponentially with the use of properly designed computing; the demand for computer information services increases exponentially as information quality increases.' The result of this self-reinforcing spiral is of course increasing dependence on computers. The more complex the system, the more it needs computers to run it; the more computers are used, the more add-ons and upgrades are necessary to run still more complex systems.

Computers are not only a colossal investment and overhead: the third largest industry in the world has grown up on the back of a product which sells itself as the means of making other industries more efficient. In practice, alongside their undoubted benefits, networks of computers offer unparalleled opportunities for both humans and other computers to screw up. This is the real risk in computing: on the human side, the insertion of a disk in the wrong drive, the accidental erasing of information and the temptation to tinker; on the technical side, random errors in processing, transmission and storage, the likelihood of which increases as system complexity grows.

Despite the voluminous and sensationalist literature on computer crime, the greatest threat to businesses in the use of IT is not electronic fiddling or industrial espionage but technical accident and human error. For a start, no sizeable software project is ever error free, and in complex systems there is no possibility of foreseeing how it will react in all possible combinations of circumstances. (Hence the controversy over the safety of the fly-by-wire A340 Airbus; a case where systems resilience is critical.) Figures gathered by the French insurance industry showed that of the £700 million computer-related losses reported in France in 1986, 44% were due to deliberate incidents; but of a total of around 36,000 reported incidents, no less than 34,000 were accidents or errors (Coopers & Lybrand, 1988).

It is true that computers offer dedicated hackers and fraudsters new opportunities for highly original trickery. Volkswagen's £170 million loss in a computer currency fraud was a warning of the potential for loss even in established systems. It is also suspected that fraud totals are substantially under-reported. But as with manufacturing automation, the key to information security is not ever more complicated technology, which may make technological deception even harder to detect, but better management. Most of the risk must be contained by simple systems design plus sensible administrative control procedures, in which case computers can yield a useful improvement in security. As Schweitzer (1986) pointed out,

> 'Employee motivation, training and awareness are critical if business information resources are to be controlled in the electronic age. Security technology is a help; but without the interest and concern of the majority of employees, trouble will eventually occur. The serious threats about which managers must be concerned have to do with mundane things like procedures, training, motivation and supervision.'

Viruses

Viruses threaten both the availability and integrity of a system, the principal difficulty being in detecting what changes the virus has made.

Viruses tend to enter the system on transferable media, such as floppy disks, containing public domain software. There has also been an increase in viruses entering software supplied from reputable sources, so that purchasing of software has to be carefully controlled.

There are now well over 1000 known viruses. The older (and less malign) viruses are more common. New 'stealth' viruses replace the last few bytes of programs to keep the size the same and modify themselves to keep check-sums the same. There is also a new class of 'armoured' virus. Many new viruses come from Italy and Bulgaria.

It is estimated that at least 60% of virus attacks are not reported; many companies see it as a standard Monday morning job to clear out last week's crop of viruses. The problem is that the job gets longer as more viruses are spread and as they become harder to remove. One response is an increasing tendency to remove exchangeable disk drives from PCs. Games are banned and programs may not be brought in or copied. Personnel are not normally permitted to take disks or personal computers home. It is not possible to check all disks coming into buildings, and legitimate use of disks brought back from training courses has been a source of viruses.

continues

continued

The operating systems of some personal computers seem to be particularly susceptible to virus infections and many have had to have virus-checking software installed. Some organizations have found it necessary to add password protection to their PCs.

There are some 35 companies in the UK working in the field of virus protection/removal (including six disassemblers and six anti-virus product companies). Viruses are often given multiple names by different companies and there is a duplication of effort in determining mechanisms to combat them. This is because there is no liaison between the companies, despite the fact that there is more than enough work for all of them. It has been suggested that a national library of viruses, with a central index and common naming, would help the industry. This is unlikely to be set up at present because of intercompany rivalry and because of lack of police manpower. However, there is a need for a central body to coordinate protection against viruses.

These considerations apply all the more with the dramatic spread throughout offices and factories of the personal computer (PC). The security problem is inherent in the nature of the beast. The attraction of the PC is its effectiveness as 'a bicycle for the mind', and not surprisingly the pioneering microcomputer companies have emphasized all-purpose ease of use as a means of selling the product. In business, however, the all-round ease of use can be a handicap without proper security measures to prevent it being exploited by those who have no business using it at all. Mainframes and minicomputers usually have hierarchical security procedures built into their operating systems to prevent users from penetrating further into the system than their rank allows. These are not at all infallible – hackers have broken into GKN's R&D database in the UK, NASA and even the Department of Defense in the USA, among many other less well protected operations – and there are great difficulties in making the security measures used by different manufacturers compatible with each other. There is no such safeguard in PCs at all, making them extremely vulnerable to both data and hardware theft. Not surprisingly, the Coopers & Lybrand (1988) report found that microcomputers formed the least satisfactory element in companies' network security provision.

Information abuse

Apart from the well-publicized cases of sabotage, computer blackmail and fraud where insiders have profitably tampered with accounting and currency routines, information abuses using the computer range from the invasion of

continues

continued

personnel records (with possible problems under the Data Protection Act) to damaging new product leaks. One company found that over a period of months a competitor appeared to have inside knowledge of its tender prices. Having resisted computers for years, journalists on the UK's new technology newspapers have quickly learned how to read not only their own files and other people's, but also those in other departments, confidential business memos and correspondence relating to trade union matters. It is hard to prove, but there has been at least one instance where journalists believe that a story was electronically 'stolen' by another paper. Ironically, editors in one office have taken to committing their most confidential thoughts to paper rather than their computer screen.

Implementation

Smooth, efficient running of a system depends on the nature of the information resource having been thoroughly understood and the company's needs catered for in the design stage. This, however, is not enough. Any consideration of the risks to business information must also take account of the special conditions applying during implementation.

Financial risks

First, there is the financial risk in the manufacturing or information systems investment. As is well documented in the financial management literature, risk and return are intimately connected. Companies striving to make bigger returns than their competitors must accept that they will incur bigger risks. This should have been dealt with by careful financial planning, including where appropriate the use of modelling and simulation techniques. The important thing is to know the extent of the risk and plan how to cope with unpleasant outcomes using damage limitation principles common in warfare, especially in navies. A useful analysis technique to compare different eventualities and decide on how much to spend on avoiding them or reducing their impact is the concept of expected value:

> Potential loss from the event concerned × Probability of the event occurring

Attack the events with the highest expected value of loss first.

On the whole, the simpler the systems, the smaller the financial risk. If integration simply consists of moves towards JIT, for example, there may be almost no financial threat in its implementation at all, although the issue of systems resilience must be duly weighed and decided. At the opposite extreme, a firm aiming towards CIM is taking a very large risk, the outcome of which may well be decided by the way in which the implementation is carried out.

Parallel or modular integration

Any programme of large-scale computer integration can be designed and undertaken in one of two ways. At one end of the spectrum, it can be viewed as a completely integrated programme in which all the parts have been designed to mesh together in a comprehensive and seamless system. Such a scheme in theory offers the highest pay-off in synergies between the interacting components and in moving at once to the new high levels of systems efficiency. But it is extremely risky to upset all the organization's tried and trusted systems at once. The risk may be worth taking if the potential prize is equally impressive: a dominant cost position in the industry or the ability to bring a range of new products to market decisively faster than rivals. The important point is that the risk should be recognized and objectively considered before the decision is made.

At the other end of the spectrum is the modular, 'islands of automation' approach, in which the programme is divided into modules or partitions to be implemented in sequence. Each module is designed, tested and put into practice before the next section is addressed. In fact, at the extreme the divided approach can be almost as risky as the comprehensive leap, since the changes lose urgency and momentum, and the gains through synergy take longer to show through. The ideal is probably somewhere in the middle of the spectrum: implementation of groups of related segments, each of which has real benefits itself and links up logically with other elements to offer still greater ones. These need not be all in one department; in fact it is usually better, because of scarce management resources if nothing else, if they span different functions. For example, CAD/CAE (computer aided engineering) might form an individual module in the programme, linking with purchasing and manufacturing engineering to provide the wider benefits of an integrated service to production.

The need to manage risk during implementation adds powerfully to the overall incentive to keep the technological element of integration/automation programmes as simple as possible. The more straightforward the solutions and the smaller the number of different computer architectures, the easier the task of integration is likely to be. Avoidable complexity should be carefully guarded against, and managers should make that criterion clear from the outset. The goal is to achieve an information management system which uses the minimum of information technology consistent with the firm's business aims, not the maximum.

Operations

The risk of information glut and overcomplexity can be avoided by care and attention at the design stage. Vulnerabilities to information loss or damage can similarly be minimized. But there is also a range of procedures and precautions which should be adopted in operations as sound information management practice. They are: physical controls, barriers such as walls, locks and guards which diminish risks by physically preventing abuse; administrative controls, for example in ensuring that people who will be handling information and computers are qualified and competent to do so; and technical or logical controls such as passwords which are built into information systems to prevent unauthorized access. In addition, companies dependent on high tech information systems should have contingency plans for dealing with disaster if the worst happens and the computer systems or communication networks break down. In practice, good security depends on a combination of all these procedures. Passwords are no use if they are never changed or removed when people leave, so they must be accompanied by administrative controls to make sure they are. Likewise, contingency plans must take account of a firm's changing physical circumstances.

Physical controls

These are the simplest and most obvious insurance against computer misdemeanour or accident. Physical protection includes putting computers in secure rooms or installations, preferably with a single entrance and exit, making them as fireproof as possible and installing burglar and fire alarms. Sounds elementary? In one recent case a firm persistently ignored the data processing manager's warnings that the computer room in a new building was inadequately protected from fire. The building went up in flames, and the company was only saved from disaster because the worried manager on his own initiative had decided to keep copies of all source programs and documentation at home. Compared with the other categories, simple physical controls are the most likely to be adequately implemented in themselves, although they are of limited utility until they are combined with administrative controls and, even more important, backed up with the training and motivation which makes their purpose self-evident. In addition, recent research has shown how easy they are to circumvent: emanations from terminals, for instance, can be picked up over a substantial range by relatively unsophisticated eavesdropping devices. Nevertheless, one estimate is that 95% of all damage to computer-based information could be avoided with a combination of commonsense physical and administrative controls.

Software theft by employees

The EC's software directive (implementation date 1 January 1993) came into being because of the increasing incidence within Europe of software theft and piracy.

Germany, with its absence of a workable law on the subject, has the highest losses in Europe, but changes occasioned by the EC directive will allow software publishers at least to take legal action to protect their copyright.

The UK already has strong copyright laws and the Federation Against Software Theft (FAST) has been instrumental in setting up, among its members, a scheme which covers software auditing, policy manuals and all the necessary materials for staff education needed to combat theft among employees. It has also encouraged firms to take action on copyright issues by including copyright clauses within employee contracts and some companies are cooperating with software companies to make unauthorized copying of software by an employee a disciplinary offence.

In 1992 the Software Publishers Association (SPA) held its annual conference in France and stated that Europe must make an effort to maximize the effectiveness of all the antipiracy organizations throughout Europe. However, it was evident that different countries are at different stages of development in terms of education and awareness of the problems and so a Europe-wide set of regulations arising from the EC directive could be some time in coming.

Administrative controls

Developed administrative controls are most likely to be lacking in small firms. It is tempting to sympathize with such a failure, since office controls are essentially bureaucratic and in themselves add little value to the manufacturing process. However, they are an indispensable insurance policy against the obvious forms of computer fraud and damage. For example, it is essential that companies take up references for all new recruits who will go anywhere near a computer, and verify them particularly thoroughly in financial and treasury operations. This sounds blindingly obvious. But with the present short supply of competent programmers and operators, a surprising number of companies fail to carry out these elementary checks. The more senior the position, the more important it is not to take the reference for granted.

Among a number of instances, an Economist Intelligence Unit special report on computer fraud quoted the case of a senior computer manager who was charged with blackmail for taking from his firm all the copies of certain key tapes and demanding a ransom. While awaiting trial, he applied for other jobs through an agency. 'He was given work as a senior computer manager by another company which did not appear to read newspapers. The company never sought a reference from his previous employer, the victim of the blackmail attempt' (Kelman, 1985). The report cites another case in which a programmer inserted logic bombs in his employer's computer code and then demanded money to remove them. When the police were called, they discovered that he had done the same thing twice before with other employers who had failed to report the matter.

The literature on computer security is extensive, and although much of it is guilty of the strategic error of putting the technology cart before the information horse, it contains useful technical detail of interest to the nontechnical reader. For the purposes of this discussion, however, some general points need to be made.

- Whatever its form, the first step is to make the security programme a formal one, backed up by management statements that it means business. Like fire procedures, it should be known and publicized.
- As a rule, information should be available on a need to know basis. This means identifying important information and classifying it according to value. All classification systems tend over time towards the excessively secretive and bureaucratic, but it is important that all levels of IT users treat their computer information with the same degree of concern that they use for paper money. The information often, after all, represents money to the right, or wrong, person. Special protective measures (see below) may be needed for the most confidential documents.
- As Schweitzer (1986) pointed out, information in a computer network is like water in plumbing: capable of being turned on or off at the turn of a tap. It cannot therefore be controlled in the same physical way as paper. To back-up specific technical controls, general measures like nondisclosure or contractual agreements are needed to secure the company's business rights to its proprietary information. The prudent handling of commercially sensitive information on computer is particularly important, since it may affect the outcome in legal cases.
- As far as possible, no single person should be the only one who understands how a process or part of a process works.
- Pay particular attention to the obvious sensitive points in a system. These include all personal and distributed computing, computers on the shop floor and anything to do with finance. In large and medium-sized firms, the auditing problems of moving towards an IT-based accounting system will have been dealt with; in smaller companies they may still be an issue.
- Electronic links between firms and their suppliers or customers are susceptible to misuse, and the casebooks record a number of ingenious intercompany collusions. Controls and regular audits are needed in accounting, order processing and wages, another fruitful area for fraud. The personnel department will also require periodic audits to make sure, for example, the

computer passwords are withdrawn when an employees leaves. There are numerous cases of ex-employees continuing to use their old company's computer systems in all manner of ways after they have left, sometimes to run small computer consultancies.

- R&D databases and records about new products contain much high-value information, particularly if the wider picture can be pieced together from the technical fragments and more general information available in 'public' files. New product information unprotected by patent or nondisclosure agreement has been and still is the basis of many a business start-up; in other cases technical details have mysteriously appeared in the business press long in advance of official publication.
- Remember that information loss through theft or deliberate leakage is still far more likely to happen through paper rather than computer systems. Whitehall moles use photocopiers rather than personal computers. Far from making physical copies redundant, of course, computers are tireless generators of paper. What happens to all those printouts when people have finished with them?

Technical controls

These are electronic security measures used to safeguard business information in line with its value. There are three main kinds.

(1) *Passwords,* which require that the user authenticate his or her identity before being allowed to use the computer. Passwords are often guessable, and for the determined electronic intruder the number crunching capabilities of the computer itself are well suited to trying out the different combinations. In addition, passwords cause problems for complex systems with thousands of users employing different equipment to run different applications on different sites. On mainframes and minicomputers, password controls for many years have been built into operating systems. Even these are not tamper-proof, but they are a great deal better than the minimal protection offered on microcomputers, where passwords only apply to applications programs and can be coaxed out of the system by anyone with a reasonable knowledge of how computers work. Passwords need to be carefully managed by administrative controls to maintain secrecy and ensure that they are regularly changed.

(2) *Access controls* limit users to certain predefined areas of the system. They also log the activities of people using the system, a characteristic which has been vital in identifying computer break-ins and catching the culprits in several cases in the past. In general, the more user-hostile the technical controls, the more users will either attempt to devise means of getting round them or minimize their use of computers, both of which cancel out any security gain. Being invisible and imposing no tedious routines on

the user, access controls are the best kind of security measure and generally effective for most routine computing (although not infallible).

(3) *Encryption,* that is, programs which scramble data so that if it is lost at least no one else can understand it. The best encryption codes are to all intents and purposes unbreakable (this also applies if you lose the key), but they obviously impose a certain overhead in terms of computer routine. Encryption is so far the province of defence establishments:

> 'Message encryption has few users in business and commerce and there does not appear to be a clear understanding of when and how to use it. Governments take national positions as to what is or is not allowed. Existing regulations are not generally known about, nor are they clear or tested. While products do exist, key management and distribution, performance constraints and economics present additional problems over and above national regulatory differences.'
>
> (Cooper & Lybrand, 1988)

Contingency plans

Murphy's famous law is that anything that can go wrong, will go wrong. Less well known is O'Reilly's law: Murphy is an incurable optimist. In 1975 the ceiling fell in on Plessey's computer centre. Plessey survived. Not so a medium-sized company, described in the EIU report, which suffered a direct hit in an aircraft crash. The crash wiped out all its computer records; without back-up or contingency plans, it was unable to continue operating (Kelman, 1985). Few companies have adequate and rehearsed contingency plans to deal with disaster of lesser mishaps to their computer networks, partly because of the multivendor nature of their systems, partly because of the complexity of networks which far outreach an individual span of control and overwhelmingly because information is not managed as a key business resource. The fact that disaster recovery has suddenly become a hot computer subject is a condemnation of the attitude to risk of both vendors and users in the past.

Contingency plans should not only cover the means of recovering IT networks in the shortest possible time. They should also include manual back-up for the minimum information needed for the company to keep operating. Taking customer orders and placing purchase orders would normally come into this category. Contingency planning for IT is impossible without a strategic view of information as a whole, since it demands a degree of information evaluation and classification. Ideally, the issue would be handled at the systems design stage, where requirements could be fundamentally simplified.

Security of information cannot be bought off the shelf. Absolute system invulnerability does not exist, nor is it ever likely to. But companies could substantially improve on current low levels of security by adhering to the simplify/integrate/automate model of manufacturing information, taking the security implications into careful account at every stage. By obliging managers to take a strategic view of the information resource as a whole, this approach helps to prevent technology infatuation or panic in which security becomes at best another costly overhead and at worst simply impossible. How many companies, for example, took into account the availability of skilled computer manpower in designing and implementing their network systems? Yet the present forseeable shortage of operators, programmers and systems analysts is directly compromising availability and integrity of information in some of the most intensive computer users, as well as substantially raising its costs. Add to this the death of security training and standards, the failure of IT vendors to provide advanced compatible solutions and other external complications. The conclusions are inescapable: companies must be responsible for their own security salvation, and the only way of reducing risk to computer systems to manageable proportions is to plan the simplicity and the safeguards in from the ground up.

Risk of consultant bankruptcy

Consultant bankruptcy is a risk not often mentioned but one of enormous importance for an organization embarking upon major IT projects with consultant help. This was highlighted in a working paper by Helen Margetts and Leslie Willcocks of the City University Business School *(Information Systems and Risk: Public Sector Studies,* January 1992*)*. The paper cites the examples of the use of consultants in the public sector and warns against the risk of consultants who designed or wrote the systems going into liquidation. It suggests that the only safeguard against this is to investigate the financial soundness of the organization before placing contracts: but warns that this can be difficult because of unpredictability of the consultancy market.

References

Coopers & Lybrand (1988). *The Security of Network Systems:* A report on behalf of the Commission of the European Communities

DTI/BSI DISC in association with the SEMA Group (1992). *User Requirements for IT Security Standards*

Kelman A. (1985). *Computer Fraud in Small Businesses,* Economist Intelligence Un
Special Report No. 194. London: Economist Intelligence Unit

Kimmerly W. (1986). Managing the risks of installing CIM. *Computerworld*, October 13

Schweitzer J. A. (1986). *Computer Crime and Business Information: A Practical Guide for Managers*. Elsevier

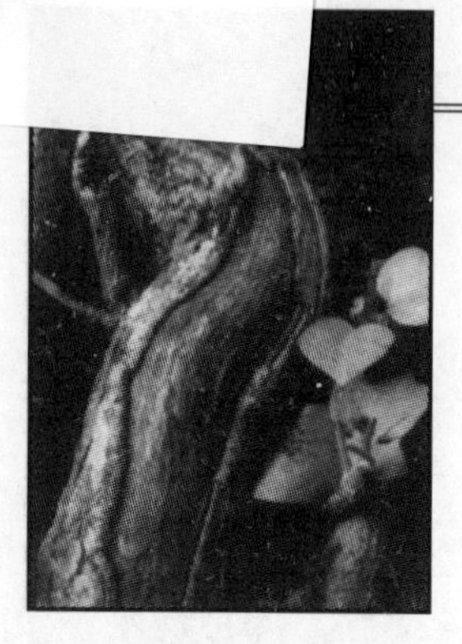

7

Value through quality

'Quality, reliability and cost are all interconnected. With enough expenditure anything can be endowed with high quality (whether through high specification and/or high conformity to specification during manufacture), and with adequate expenditure almost anything can be made to be very reliable. It follows then, that a company provide a product or a service at different quality levels, each of which necessitates a different price. There is no single level of quality: nor is there an absolute quality level. Nothing will be perfect, no matter how much it costs. In general, costs rise steeply for increasing quality, but beyond a certain level, value to the customer increases more slowly. Thus, it is possible to identify, notionally at least, a point at which the difference of value and cost is maximised' (Wild, 1991).

Quality in manufacturing has always been sought in most parts of the world – the nineteenth century was probably the zenith of quality manufacturing for relatively low overheads and we shall not see its like again. The huge economic changes that took place globally after 1945 led to a period of reassessment of business techniques and practices, and the birth of the management consultant. One in particular, W. E. Deming, was the first consultant in the 1950s to write about the concept of total quality control (TQC), and one of the few westerners to influence Japanese culture, since they took and made it the cornerstone of their post-war management techniques.

Deming's philosophy incorporates some of the following:

- A commitment by senior management to the achievement of quality;
- Develop a long-term strategy for improving quality;
- Invest in high-quality materials;

- Invest in high-quality staff and train people properly;
- Do not sacrifice quality for greater output;
- Encourage the participation of all employees in the search for greater quality (quality circles);
- Educate management to accept suggestions from the workforce;
- Educate management to communicate quality objectives to the workforce;
- Improve supervision.

His philosophy put the emphasis firmly on 'control' and it was this point that most appealed to the Japanese culture. Curiously enough, western companies chose to move the emphasis to quality 'assurance', which sounded more flexible but was in fact more rigid, since it meant following a set of laid-down procedures which did not build in the elements of education and motivation. Early failures of western companies to adopt, piecemeal, some of the highly developed management techniques has led to a seemingly universal decision to go for total quality management (TQM), which embraces not only the central organization but its suppliers and contractors also.

The Computer Services Association (1991) endows TQM with the following attributes:

- The approach is management-led and management is visibly committed to improvement.
- A customer-orientation permeates the entire organization, the needs and requirements of both internal and external customers are sought, and the level of satisfaction with the services or products becomes the basis of continual improvement efforts.
- Suppliers are managed for the benefit of the organization's customers.
- Teamwork involving people at all levels is seen as a key to improving processes and services.
- Quality management and improvement education are provided at all levels of the organization, emphasizing 'prevention' rather than 'detection'.
- Accountability for quality improvements is tied to manager's performance evaluations.
- Recognition and incentive programmes are established throughout the organization, are targeted at service improvement efforts and are used creatively to encourage involvement.
- Quality measures are established and high standards are set for quality service delivery in all areas.
- People are constantly encouraged to participate in the improvement of quality.

In the TQM process everything is measured – product, service or intangible benefits.

Quality controls in manufacturing

The *right first time* concept describes a production process philosophy whereby a number of factors are combined to ensure a high quality production process. It requires detailed attention to production tasks, inputs of high quality material at the beginning of the production process and a continuous monitoring of process to anticipate problems before they occur. Right first time requires meticulous planning of production technology and thorough training of production workers to achieve this level of quality.

Zero defects is a concept closely related to right first time and refers to an acceptable level of quality that is sought. Right first time manufacturing processes are thought to be the ideal way to achieve zero defects but it can also be achieved through non-technological means, that is, by continuous sampling, inspection and checks. Relying upon manual checks obviously opens up the process to greater chances of error.

Statistical process control (SPC) involves regular sampling and immediate analyses of variables taken. The resulting statistics should give a picture of any deviation from the required result and adjustments in the production process can be made. SPC relies upon setting up a pre-control chart prior to a production process starting, charts to be used during the running life of a process and CUSUM (cumulative sum) charts to show if a worrying trend in sample results is developing. Originally a manual control system, it can now be achieved through a computer program.

Quality buzzwords

Companies innocent of quality often find themselves unable to understand the language of quality management. But, while the vocabulary is large, the meanings are simple.

Total quality management or TQM connotes a holistic management system, as outlined above, and goes beyond quality assurance to embrace everything that a company does. *Benchmarking* describes the way companies try to identify internal and external standards as 'excellent' and use them as standards to be emulated. *Ownership* is a term commonly said to reflect the sense in which quality is a collective responsibility, not a functional one. *Cost of quality* is (confusingly) a definition of the cost to a company of not achieving quality – cost of lost customers, and of identifying and correcting errors, minus the cost of fostering quality.

More technical terms come under the heading of 'translation tools', which include any formalized method of turning customer expectations into design and manufacturing requirements. The most commonly used are:

- *Avoidable input analysis:* a contribution to the quality lexicon from Amex, it is a method of turning customer enquiries into product and service modifications with the aim of eliminating the need for the enquiry.
- *Pareto analysis* is a simple system to identify the 'vital few causes of a problem, and to direct attention to these, rather than waste time on the trivial many'. It is based on the maxim that 20% of the problems cause 80% of the trouble.
- *Paired comparison* is a means of prioritizing nonquantifiable causes or issues. It establishes priorities from a list of possible causes of a problem by comparing causes in pairs.
- *Ishikawa diagrams* are used to analyse the causes and effects of a problem. They are flow type diagrams, which group all the causes and effects of a problem under the '4M' heading – materials, machines, manpower and method.
- *Taguchi experiments* are used to examine complex processes with many variables and to establish within a relatively limited number of trials the main control factors and their optimal levels.
- *Hoshin plans* are a goal setting and strategic business tool. The first stage is to identify an owner for a process or a product. Then a strictly measured and meaningful goal is established. A plan typically lasts for a year, but the Hoshin plan is reviewed every quarter.

Companies that tread the quality management path will realize that all the existing plans, systems and techniques are as nothing if the message cannot be got over to the organization as a whole – managers as well as workforce. Therefore, experts advise tailoring a quality management approach which integrates cleanly with the best elements of an organization and attempts to eliminate the worst aspect of the same.

Change is not always welcome

The quality message is not necessarily a comforting one. Quality is about change, and change is frequently not welcome. Perhaps the best writer on quality, J. M. Juran, echoed the 3M sentiment when he said that management is responsible for 85% of product failures. Juran believed that there must be a 'management breakthrough' where the flawed routine is broken and a regime of continual change installed.

Case study: 3M

3M's quality approach is labelled by them – managing total quality. Its success is not in doubt. The company was set up in 1922 as Minnesota Mining and Manufacturing, with the intention of marketing an industrial minerals deposit that turned out not to exist, and now manufactures 60,000 products ranging from Scotch tape to diagnostic laser imagers. A third of its $13 billion sales come from products introduced in the last five years. But 3M has described its business not in terms of products but in terms of quality:

> 'Quality is viewed not as additional work but as the framework and methodology from which we approach our jobs. Quality becomes the focus used to achieve our goals – why we are in business.'

In other words, quality has become the corporate culture at 3M. But this is 3M culture – not a transplanted Japanese culture. So it is not surprising that the US company has found a recognizably North American way of expressing its conception of the quality company. The company speaks of 'transferring the ownership of quality' to the whole workforce, of 'empowering' the individual employee through the quality process. This participation in the quality ideal is achieved through understanding and implementing the company's own exposition of how quality can drive an organization – 3M chooses to call this exposition the 'five essentials' and the 'eight elements'.

This handful of catchphrases may sound bizarre to the uninitiated. But what 3M is trying to achieve is a quality philosophy expressed on its own terms – terms that will make sense in Minnesota. So this quality language is designed to strike a chord with dreamers of the traditional American Dream of universal success, where every person becomes a king – here seen as the 'empowering' of the individual employee. The language will also be familiar to advocates of more prosaic systems of 'employee participation' in the fortunes of the company (although capitalists will be relieved to hear that 'transferring the ownership of quality' does not involve any dilution of their stockholdings). In the quality language, these essentially traditional themes are spiced with another, more exotic flavour: the 'five essentials' and the 'eight elements' round off the ideas with a mystical, almost transcendent note. This is a note to which management thinkers are rather susceptible – perhaps it conjures up the inscrutable power of the Japanese management machine. But it all adds up to a potent new use of ideas that the employee already knows.

The heart of 3M's approach to quality is the idea that quality performance is an organizational issue. The company has said:

> 'Only 15% of all errors are resulting from the workforce; 85% are attributable to management and its systems. Errors often result because those who work within the systems misinterpret them and because management is unwilling to solicit or listen to new ideas.'

The routes for achieving a quality culture are many, but the common factor is the need to be able to absorb change. Quality is not static, it is dynamic and challenging.

When a company embarks on quality, said McKinsey consultant Michael Graham, 'they take on board something that is dangerous. The status quo is put under threat, and the outcome has to be unpredictable'. According to John Pike at the UK's Danbury Park Management Centre, total quality management is about continuous improvement. It involves employees in the process of identifying and solving problems. Tony Barnes of Crosby Associates put the concept more graphically when he said that Philip Crosby (author of *Quality is Free*) considered a quality initiative revolutionary because it brings about a shakeup in the organization that can cause considerable trauma and stress.

Does technology let the side down?

It used to be thought that machines were better than people because you could control them more effectively. However, more and more companies are finding that the struggle to master IT and to absorb it into the business culture, or to change the organization to fit in with a computer integrated manufacturing scenario, is a continuous struggle and, in some cases, self-defeating.

> 'Automating all the routine work, integrating all our control and communications systems, running the business through decision support systems at every level – it's a massive job. We just don't have enough change agents to make it happen'

bemoaned one director interviewed by Price Waterhouse (Grindley, 1991). One of the dangers of change is loss of control and, in manufacturing, say the experts, quality is all about control.

The British Standards Institute's BS 4778 *Quality Assurance* states that

> 'Quality control is that aspect of quality assurance which concerns the practical means of securing product quality as set out in the specification.'

Wild (1991) defines control of quality in manufacturing as being a three-stage process: control of inputs, control of process and control of outputs.

Control of inputs means applying quality standards before the production process begins, that is, requiring specific levels of quality from suppliers of raw materials or components; setting standards for suppliers of services; and equally stringent standards for subcontractors. The retail chain Marks & Spencer were in the forefront of quality control at least 20 years ago. Not manufacturers themselves, they nevertheless demanded a level of quality from suppliers that many manufacturers could not match. Specification manuals for the 'rag trade' for example, were dauntingly thick – buttonholes had to be so many millimetres, no more no less; yarn used had to conform to high strength specifications and so on. Contracts to supply M & S were hard won only after rigorous standards of manufacture and supply were met.

Control of process, as discussed before in this chapter, requires constant checks, inspections, analyses and modifications/corrections to be made during production, using established techniques. As Hill (1985) pointed out, the task and responsibility for quality of manufacture depends largely upon the type of manufacturing process. There is a distinct difference between project and jobbing, continuous process and batch and line.

Control of outputs means a final inspection of goods produced before they reach the consumer. This can be in the form of representative sampling, backed up by quality assurance techniques or, as in the case of high value, labour intensive products, a comprehensive check of each item.

Case study: LMG Smith Brothers

LMG is part of the $1.2 billion Lawson Mardon Group, a Canadian-based international packaging company which employs 7000 in six countries.

LMG's plant workload is particularly heavy at peak times and the company felt that batch routines and response times were too slow. Recurring problems with system performance caused the company to rethink its IT strategy and find a system that matched its needs.

The needs were for a system that could provide accurate real-time reports during the production process and run batch routines without hindering system performance. LMG also had its own in-house developed Pick system for monitoring the whole manufacturing operation and had chosen to adopt an open systems UNIX strategy servicing 150 users across four sites.

The company demanded fault tolerance and chose a Sequoia Series 400. 'We looked at some of the high resilience systems available but these lost a considerable amount of time during back-up after a fault', said a company spokesman.

Thus far, the Sequoia has provided virtually uninterrupted running and is linked into the supplier's online fault monitoring system in the US, which checks for any malfunctions and despatches parts to the company for faults that it was not aware of.

Waters (1991) suggested the times in the production process when inspections are most useful are:

- At material suppliers' plants during their processing,
- On arrival at the plant (including all materials from suppliers),
- At regular intervals during the process,
- Before high cost operations,
- Before irreversible operations (like firing pottery),
- Before operations which might hide defects (like painting),
- When production is complete,
- Before shipping from the plant.

The theory and the will to succeed with quality control measures is there in many companies, but does the technology match up to the reliability of the human TQM resource? Apparently not, since we have only just entered the era of 'fault-free' technology, as *Computer Weekly* (1992) described it, but traditionally conservative mainframe suppliers prefer to describe their machines as 'fault tolerant', with claims of as high as 99% reliability.

Technology failure can cost industry huge amounts of money in lost production and surveys have shown that it is among one of the greatest concerns of organizations aiming towards CIM. Hardware manufacturers are cognizant of this fact and several, such as Sequoia, have set up national or even global networks which monitor all installations around the clock for early detection of faults. Quality control starts with suppliers. The organizations that demand quality from hardware manufacturers have begun to impose their standards on the computer industry. Software, however, is still a cause for concern.

In 1988, a review of software quality standards by Price Waterhouse on behalf of the UK Department of Trade and Industry estimated that UK users and suppliers suffered software failure costs of over £500 million a year. This alarming figure, however, only represented domestically produced, marketed software. If imported software and software produced in-house had been included the failure costs would have been much higher. Software failure also incurs indirect costs, which can be substantial but difficult to quantify.

The review suggested that a quality system could help to improve software quality and reduce failure costs. (Quality systems provide a structure that facilitates the creation and monitoring of quality. All aspects of software production and maintenance can be addressed by a quality system, including development procedures, quality control and quality assurance (QA)). But a survey for the same government department conducted four years later by CSC Index (1992) showed that only one-third of the companies interviewed had introduced quality programmes specifically addressing business systems. Approximately half of those companies had set up quality assurance units to support the development of business systems. When asked, these companies identified 'carrying out QA audits of project deliverables' as their prime activity while 'assessing the extent to which users' needs will be met' and assessing the adequacy of user involvement' were the two lowest rated activities on a list of 13.

It would appear, therefore, that the key objective of quality programmes is not being addressed directly by the QA units; neither are user satisfaction and levels of waste better for organizations with such units. The survey did not identify the causes for this, however, but it suggested two main reasons:

- The introduction of 'quality units' raises the expectation of users. Therefore, improving user satisfaction requires more than small incremental improvements in quality procedures.
- The main activity of existing quality units is to review what has been completed rather than to concentrate on building quality into the systems. While conforming to specifications and standards is important, getting the specification right is more so.

The survey then goes on to comment that quality assurance units are more often staffed by technicians whose aim is to improve the technical quality of the systems rather than to evaluate quality as defined within a business context. So we come back once more to the IT/business culture gap outlined in Chapter 5.

One consultant believes that IT managers need to use software metrics to get a better picture of performance than traditional tools if they are to convince their superiors that systems are doing their job. By careful analysis of systems strategy, management and staff capability and user satisfaction, among other key areas that need to be considered in the assessment process, it is possible to generate meaningful indicators, or metrics, which can be used to drive a process of continuous improvement (Stalley, 1992).

Case study: Xerox 1983 to 1987

In those five years Xerox Corporation cut manufacturing costs by 20% despite inflation, reduced product development lead time by 60%, boosted revenues per employee in its major business unit by 20% despite lower prices, and became the first US company to regain market share in an area targeted by the Japanese, without the aid of tariffs or protection.

Xerox ascribed the improvement not to automation and factory hardware, but to an attitude of mind. Like a growing number of western companies in the last decade, Xerox had discovered the productive power of commitment to quality.

continues

continued

The revelation came from the East. Mirroring the global performance of the parent, Fuji Xerox in Japan had been a runaway success whose performance dipped alarmingly when it ran up against its first determined competition in the mid-1970s. To dig its way out of the trough, Fuji Xerox launched New Xerox Movement, a total quality programme which sought to alter the entire company culture. To symbolize the new commitment, Fuji Xerox inscribed on its list of objectives the winning of the Deming Prize, Japan's leading award for a company-wide quality process.

New Xerox Movement was an unqualified success. In 1980 Fuji Xerox won the Deming Prize, regaining its financial momentum in the process.

Impressed, Xerox decided to implement a corporate quality strategy over the whole group. A new policy statement proclaimed:

> 'Xerox is a quality company. Quality is the basic business principle for Xerox. Quality means providing our external and internal customers with innovative products and services that fully satisfy their requirements. Quality improvement is the job of every Xerox employee.'

To translate the principle into practice, a quality implementation team devised Leadership Through Quality, a plan which laid down strategic goals and developed five levers for change.

- *Standards and measurements* to provide people with systematic ways of assessing their work, solving problems and improving quality.
- *Recognition and reward* to ensure that staff are motivated to practise quality behaviour.
- *Communication* as a means of informing staff of corporate objectives and priorities.
- *Training* to teach all employees the philosophy and the techniques of quality improvement.
- *Management behaviour and actions* to set the example of rigorous implementation of Leadership Through Quality right through the company.

The impressive results were, according to John E, Kelsch, Xerox director of quality,

> 'merely the tip of the iceberg. Estimates are that the "cost of quality" for a large corporation like Xerox is about 25% of revenue. For Xerox, that translates into more than $2 billion. That's an opportunity that Xerox people around the world are slowly but surely chipping away at.'

References

Computer Services Association (1991). *Total Quality Management*. Executive Summary

Cordy F. and Mansell-Lewis E. (1992). Enter the fault-free zone. *Computer Weekly,* September

CSC Index/DTI (1992). *Key Issues Affecting Quality in Information Systems*. London: HMSO

DTI/Price Waterhouse (1988). *Software Quality Standards: the costs and benefits*

Grindley K. (1991). *Managing IT at Board Level.* London: Price Waterhouse/Pitman

Hill T. (1985). *Manufacturing Strategy: the strategic management of the manufacturing function.* Macmillan

Stalley B. (1992). A wider view can show the value of software. *Computer Weekly*, August

Waters C. D. J. (1991). *An Introduction to Operations Management.* Wokingham: Addison-Wesley

Wild R. (1991). *Essentials of Production and Operations Management*. Cassell

8

Instigating, anticipating and managing change

Process innovation

In the 1990s industry is at the same time tied by the economic constraints of the recession and driven by the need to compete effectively. The dependence upon IT has meant that many companies have had to make radical changes to survive – they have had to 're-engineer' their business functions.

Just when they thought they had mastered the philosophy of TQM, along came process innovation (PI) to confuse them completely. It is also sometimes, but not always, known as business re-engineering. In June 1992 consultants Ernst & Young surveyed 37 companies in *The Times* Top 100 and found 21 different terms used to describe business re-engineering. When asked to define the concept, there was even more variance – some respondents' definitions are given below:

> 'The analysis of the business process with the objective of restructuring or re-engineering to improve business performance',
>
> 'a type of technology, systems and management philosophy of examining, doing and seeing if there is a better way',
>
> 'the developing and streamlining of processes to achieve one's business objectives and to establish the attitudes and culture to enable these objectives to be achieved',
>
> 'an attitude of mind within a company, translated into refining systems and processes to enable it to determine its objectives'.

Ernst & Young's definition of process innovation is:

> 'A way of designing the organisation, and systems that run within the organisation, to give the best effect to the organisation's capability and business aims.'
>
> (Ernst & Young, 1992)

Many of the companies surveyed believed that PI was virtually inseparable from TQM but most agreed that the difference between the two was that TQM concentrated on raising the quality of existing functions, whereas PI helped to design new functions.

One writer claimed in an article in *Computer Weekly* that, with information technology often playing a leading role, business re-engineering can achieve radical benefits in areas like revenue generation, cost base, product quality, customer service, and lead time, in some cases saving twice the cost of implementation within a year (Haughton, 1992). But, the article expands upon the theme –

> 'Effective process innovation requires a creative approach, embracing not just IT and business processes, but products and services, the organisation structure and employee skills. Process innovation is essentially a creative activity. You can't look it up in a text book, you have to sit down with a blank sheet of paper and think.'
>
> (Haughton, 1992)

You may not be able to refer to textbooks for recommended techniques for PI/business re-engineering but several consultants have developed simulation tools to model processes dynamically in both service and manufacturing indus- tries. Coopers & Lybrand, for example, developed one called SPARKS whic uses a dynamic modelling approach. The activities in the process flow ar 'brought to life' by simulating how the business performs in response different levels or patterns of work, and different levels of resource availabili The user can see the results of the simulation by watching the simulatio action and by measuring key statistics such as:

- Resource utilizations
- Costs of activities
- Timing of activities
- The location and size of queues or backlogs.

In this way the most complex processes can be analysed and performan assessed. Building up a model like this begins with analysing the curre processes, and is augmented with specific data collection exercises. Ideas f re-engineering may be generated through many methods and they are the tested, individually or *en masse* by modifying the model and quantifying th effects. As a result of the dynamic analysis provided by such models it i possible to test new ideas in an experimental way and business re-engineerin becomes a less risky commitment because the impact of change can be predicted and understood in advance.

Case study: Rank Xerox

In 1991 Rank Xerox UK undertook a totally comprehensive business re-engineering programme. It initiated a series of projects – each with a life span of two years – to examine and change, if necessary, all its major business processes, from taking orders to equipment delivery.

It mapped out all the company's processes using its own business process analyser software and the model gave a clear picture, through flow charts showing all the major processes and responsibilities. It allowed the company to understand all the links between processes and gave it a route map for re-engineering. The company then reviewed its current procedures, redesigned them and developed new information systems to support the re-engineered processes.

The first project within the planned re-engineering process was the area of order management. The cost of implementing new solutions was some £6 million, but Rank Xerox expected that the project would pay for itself within the roll-out period and eventually save the company over £11 million a year.

'Process re-engineering [as Rank Xerox have named it] is a vision-led, fundamental redesign looking for the optimum combination of people, process, information and technology', said the manager of business process development. 'And it meant a new emphasis in the IT environment. Information now has to be shared and available to all who need it. Self-managed work teams demand up-to-date information in a simple readable form. We still have a mainframe-based, centralised approach at the moment but the architecture could change.'

Anticipating change

This is perhaps the trickiest area of managing a business in today's increasingly fluid industrial environment. And the involvement with or dependence upon IT only serves to aggravate the problem because technology itself changes so rapidly.

However, IT is just one aspect of the pressures to change which exert themselves on businesses. The global market continuously creates more and more pressure to compete effectively against foreign incursion into what were once protected domestic markets. The success of the technology-based Far Eastern countries almost pales into insignificance besides the market-grabbing climate of the European marketplace and developments in Eastern Europe. Constantly changing market conditions everywhere mean that business have to use their IT to effectively anticipate future trends and this can only be done by gathering more and more information to facilitate decision-making.

Online databases

Armed with a PC, some communications software and a direct telephone line, any manager can tap into a huge range of information:

Press databases

The full text or precis of articles and news items appearing in domestic and foreign newspapers.

Company listings

These databases can be used to produce lists or labels for mailshots.

Company reports and accounts

These can be used to make a reasonable analysis of a competitor's or supplier's strength financially and in the marketplace.

Market research information

Most of the databases offer online access of printed reports on consumer and other markets.

Management information

A number of databases offer precis of articles in domestic and international management journals.

Real time databases

These are typically offered by the stock exchanges, for up-to-the-minute market movements.

Telephone information services

These services are typically offered by national newspapers on various business topics.

Television information services

Databases are run by the various television companies, which can be accessed by anyone with an appropriate set. By using a specially adapted phone line and tv or PC with appropriate software it is possible to link into a videotex system.

CD-ROM databases

Compact disc read only memory (CD-ROM): A CD which stores huge amounts of information and is purchased on a one-off basis and accessed via a PC with CD-ROM drive.

Fax databases

Fax databases technology is in its infancy but is another method of accessing market information through the fax system.

The major change, among others, in technology's role in the business is that most manufacturing companies are now IT-dependent (most current surveys show a huge rise in the number of companies who feel that they can no longer function without IT). This in itself creates a climate of anxiety which is added to by the need then to invest in newer and more dramatic technology to keep up with competitors. This change in the business environment has its necessary impact upon corporate strategies. The companies that fail are the ones that become obsessed with managing the machines and forget about the people. The human element is still the governing force within industry, of course.

Technology-driven companies have to invest more time and money into training personnel which, because the technology changes frequently, must be a continuous process. This in turn raises the value of the personnel to the company and their own perception of their worth. Good labour relations is about anticipating, at the outset of technology investment, the problems that are going to arise as people struggle through their learning processes and the change in relationships within a company that are going to occur as a result of the acquisition of new skills.

There are major changes that have taken place in customer demands. The proliferation of manufactured goods during the periods of economic stability led to high consumer expectations. They want choice, they want quality and above all, they want it *now*. Manufacturers have had to add to their product ranges, introduce tighter quality measures and speed up delivery times in order to keep a foothold in the marketplace. Anticipation of future trends in customer requirements has become an industry in itself. Using IT to identify customer expectations and market developments has become the norm.

So, the initial analysis of IT needs when a company is developing its technology base, and the reassessment of needs when a company undergoes 're-engineering' has to build in the element of being able to use technology to anticipate change in the internal and external environment of the factory as well as driving the company's core activities.

Moving towards the information-based organization

There are two aspects to the management of manufacturing change. The first is project implementation to achieve the numbers in the strategy. The second goes beyond implementation of the programme to alter the nature of the organization itself.

Beyond the numbers, the goal of a manufacturing strategy is to make production so highly reliable that it changes from corporate stumbling block to competitive strength. The management of production then changes too. Instead

Case study: using credit card information to analyse the marketplace

A manufacturer of audio cassettes used the information gathered from credit card purchases in filling station shops to place his products correctly. By analysing the spending power (how much was spent at each transaction), types of purchases (whether leisure products were bought as well as motoring products and, by the amount and type of fuel bought, the size and age of car) and how frequently purchases were made, the manufacturer was then able to select the right types and quantities of audio cassettes to place in particular outlets in specific regions. Thus the turnover of sales was increased by approximately 35%.

of constantly fighting operational fires, top managers can take time to plan strategically and 'fine tune' the use of manufacturing capacity for market advantage. Japanese firms expect manufacturing managers to spend 80% of their time, operatives at least 10%, on process enhancement.

Manufacturing management itself becomes a process of never-ending self-improvement in which the strategic numbers and individual projects are the milestones on the way. With this change, it takes on many of the characteristics of a support function. Within the factory, automation increasingly destroys the distinctions between direct and indirect labour. In plants where 'operators' spend most of their time collecting data, carrying out preventive maintenance and working on projects to cut set-up times, the category of direct labour no longer exists. All operators become indirects, just as all manufacturing people become service oriented.

Continuous improvement drives all world-class manufacturing; without it, it would not be world class very long. At the heart of continuous self-improvement are measurement and feedback, integrated into a logical overall information system where information is 'pulled' as required on a need to know basis. In effect, information becomes the company's organizing principle. Factory staff and managers manage an information flow, not machines. One way of describing the difference between just-in-time and conventional production is to talk about information flows. In a conventional factory, information lines are long, sometimes with no feedback loops at all, and radiate from a centre which typically carries out all scheduling tasks. In a JIT plant, most information travels in short loops and rarely goes anywhere near the centre. The maximum of scheduling is done locally by the production system itself. Conceptually and practically, this information-based organization is different from the manufacturing outfits of the past. Particularly when combined with automation, it requires different organization, different skills and different attitudes.

The role of senior management

Strategic purpose

To move beyond project implementation to the stage where improvement becomes self-sustaining demands a period of intensive change at almost every level. In this, top management's role is vital. The most crucial precondition for successful implementation of change, rarely stated, is for senior people thoroughly to understand the reasons for and nature of the transformation. ICL began a reorganization in the 1980s by sending all its directors on a week's course on the implications of change, followed at regular intervals by all the other layers of senior management. Such thoroughgoing preparation is unfortunately rare. Unwavering top management commitment and leadership based on strategic understanding are a *sine qua non* for the successful transition to a new style manufacturing organization.

These qualities become even more important as the know-how underpinning the operation fragments. No senior manager can master all the specialisms of the modern factory, each with its own disciplines and accumulated skills, from data processing and CAD to customer service. As Peter Drucker points out, with its increasing reliance on specialists or 'soloists', the evolving information-based company bears a closer resemblance to a symphony orchestra, football team or hospital than the military-style command structure of the past. This puts a premium on everyone knowing how to read the score, and on a perfectionist conductor using every trick to extract the best performance of it. 'It is information rather than authority that enables [conductor and players] mutually to support each other' (Drucker, 1986).

All specialists tend to build walls round their domains which are hard for the layperson to break through. This applies particularly strongly to data processing and computing. Whatever the degree of manufacturing automation, the enthusiastic cooperation of systems people is essential in tailoring systems to the real needs of data providers and consumers, and making them simple enough to be understood and routinely used. The best laid plans founder on techno-fear, techno-ignorance or plain techno-impatience. Everyone knows companies where expensive workstations and computer terminals gather dust in corners as designers and managers revert to friendlier pencils and paper. If this is the case with simple stand-alone applications like CAD or personal productivity tools, the danger of alienation and resistance is much greater in the case of complex integrated systems. There is therefore a special onus on managers to enlist computer staff as evangelists of the strategic (as opposed to narrowly DP) aims, and to use them as part of multidisciplinary teams. Conversely, systems people need to be thoroughly aware of the vital importance of their triple role in enabling, demystifying and teaching the operation of the information network. Few currently meet this challenge.

Communication

Communication is the means of providing strategic purpose and the second component of top management's role as change agent. This is not a question of company newspapers or exhortations on notice boards but of full employee involvement, and as such needs a chapter to itself. Here it is enough to underline that although systems make things possible, people make things happen. Humans make or break systems rather than the other way round, and most observers accept that more attempts to implement advanced technology founder on the human factor than on any other. For all the shiny new robots, AGVs and computers, factories for the time being are still man–machine systems in which the human interaction is primordial. In the UK, Ford's two-week strike in February 1988 was a salutary reminder that the more integrated and streamlined a company becomes, the greater the need to carry the work-force along with the process of change. That is top management's job.

Challenging tradition

To make change permanent, managers must devote time and effort to challenging the traditions of the past. This sounds easier than it is, since the traditions in question are almost always of their own creation. 'What is wrong with our factories,' declared Wickham Skinner, 'is that markets and machines have changed while our management routines have not.' He went on: 'We are not thinking hard enough to fit form to function. Before we should dream of instituting CIM – and we should – we must consider how the world of business is moving: fast technological change; shorter runs of more products; short cycles of new product development; higher capital intensity; more need for interested, adaptable, educated workers; and more cooperation across function boundaries. The list goes on: more stringent demands for quality; longer run planning; more need to deal with ambiguity' (Skinner, 1988).

Case study: GKN Hardy Spicer

GKN Hardy Spicer, a UK manufacturer of constant velocity joints, combined a £30 million investment in advanced manufacturing technology with a comprehensive organizational change, in order to reduce manufacturing costs and effectively compete in the 1990s.

The previous traditional work structure was replaced by an integrated system whereby system technicians enjoyed a new multiskilled role, performing all the production tasks, including systems supervision, frontline maintenance, inspection and loading. This, said the company, enabled the retrained, skilled and better motivated employees to speed throughput by giving close attention to detail without slowing down response times.

This is a weighty agenda for self-criticism. Many of the larger issues will have surfaced during the simplify/integrate phases of strategy formulation. At plant level, the change process should throw up a series of further searching questions about every aspect of business activity. Ask yourself: 'If we could start up business afresh on a greenfield site, how many of our traditional ways of doing things – procedures, processes, work practices – would we keep?' For instance:

- *Pay systems*. Traditional shop floor incentive systems do not work in a JIT environment where overproduction is as bad as underproduction, and where in any case operatives are moving away from a direct towards an indirect role. Can your current payment system or incentive scheme cope with motivating people not to do productive work if there is no current demand for the finished product?
- *Middle management*. What happens to middle managers? In the old organization, their role was chiefly one of selecting, reformatting and relaying information up or down. The responsive information-based company needs a much flatter organization than in the past, and the slimming operation will affect middle management ranks more than any other.
- *Customer/supplier relations*. Traditional arm's length relations with customers and suppliers cannot survive the move to an integrated view of business. Do customers want JIT deliveries, and does marketing understand what that means for the manufacturing process? Does purchasing have the skills to switch from playing off a large number of suppliers against each other to helping a much smaller number to improve delivery and quality performance? Such questions undermine traditional assump-

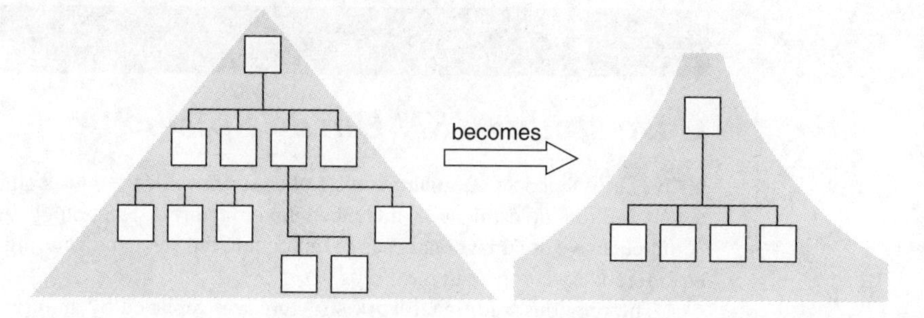

Figure 8.1 Responsive information based company. *Source:* Ingersoll Engineers.

tions of the corporation as a self-contained independent entity. A few companies, like the UK retailer Marks & Spencer, Ford and IBM, have been notably successful in upgrading suppliers over time to meet ever tougher quality standards. General Motors now puts out a manual for suppliers called *Targets for Excellence* and has renamed its purchasing department the 'supplier development' department. Other companies are following suit as they realize that, where 60–70% of a product's cost is materials, a large part of their own success depends on the quality of their supplier base.

- *Line/staff*. It is easy to see how line/staff distinctions blur on the shop floor. The other half of the equation is that 'staff' must become more 'line'. How do you get product and manufacturing engineers to make the transition? Hewlett Packard's answer is to locate them physically on the factory floor: in the same place as the problems they have to solve. A Danish company has taken the less revolutionary step of making sure that design engineers have to enter or leave work via the shop floor making the products for which they are responsible. If there are problems, they are forced to confront them at least twice a day.
- *The office*. Now that manufacturing lead times and WIP costs have been cut, what about 'invisible inventory': the cash locked up in long-winded order entry, design and ordering at the premanufacturing stage, delivery and collecting receivables at the other end of the cycle? In most companies the white collar areas are as wasteful as the factory, often even worse. Of the two months it might take to deliver a mass-produced car to a customer in Europe, just two days are spent making it. The rest of the time is taken up with order processing and delivering. Advanced companies are now applying JIT and quality concepts to their offices with significant results.

This kind of questioning should continue all through the implementation period and beyond. There is no room for complacency in a business that wants to remain internationally competitive. Although there will be discrete programmes for implementing components of AMT, the goal is not a hardware installation (FMS or CIM) but a process of manufacturing improvement as a way of life. The goalposts are always being moved further on. Accordingly all solutions are temporary. At JCB, manufacturing managers joke about putting machine tools on wheels so that they can be shifted even more frequently than they are at present.

Apart from the predicted gains, alert companies often point to completely unexpected benefits from their constant reappraisals. Edson Gaylord, Chairman of Ingersoll Milling Machine Co., confessed:

> 'When we look back at the good things we achieved along the way, it is no exaggeration to say that the most important improvements were frequently unpredicted or underestimated. We simply did not know enough to see what could be achieved.'
>
> (Ingersoll, 1985)

Another US company adopted JIT in order to cut lead times and found it was being paid for the finished article before it settled with its supplier of raw materials. The oil pump manufacturer featured in Figure 8.2 discovered $2 million in undocumented, unexplainable cost cuts. As the learning process continues, the unexpected will become less common. But the last few years have taught that it is unwise to assume anything is permanent. No one has been caught out by overestimating the possibilities of change.

Setting targets

The fourth role of top management in change is to set targets and priorities, and monitor progress against them. Overall business targets should be ambitious. Exacting demands oblige planners and specialists to question every aspect of operations, some of which may have gone unchallenged for years. In most old style manufacturing plants, overall system efficiency is low enough to permit order of magnitude improvements in several areas.

Detailed priorities vary between industries and companies. But it is important to treat the individual project as the first stage of the process of continuous improvement. For the sake of both larger and smaller goals, it is wise to concentrate the first wave of change where it has the most dramatic and visible effect at least cost. This is likely to mean the areas of simplification and integration where initial improvements do not call for large capital outlays. Consider a manufacturer of electromechanical assemblies. By reorganizing on JIT lines and spending just £250,000, this firm cut throughput time from 30 days to five hours, raised inventory turn from four to 22 and reduced rework from 20% to 1%, all in the space of four months. Taking their cue from cases like these, many people now believe that all major implementation projects

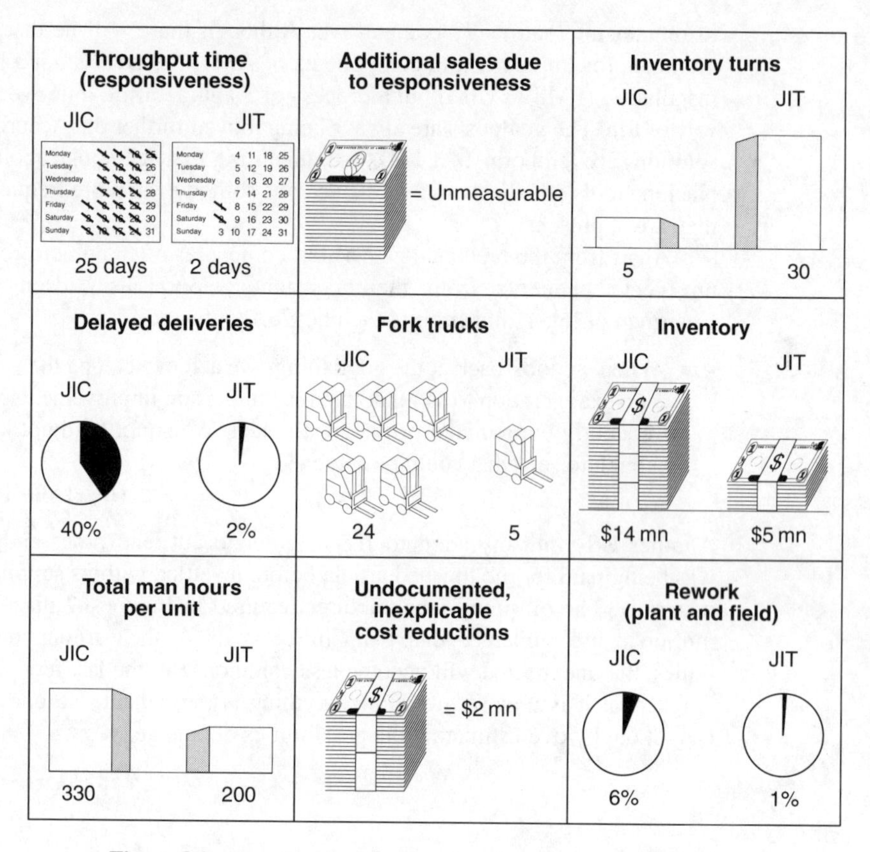

Figure 8.2 Benefits of converting from just-in-case to just-in-time: an oil pump manufacturer. *Source:* Ingersoll Engineers.

should begin with a pilot scheme which is bound to succeed quickly, against fairly modest business targets. The motivation and confidence thus created are important weapons in tackling subsequent, more challenging tasks.

Experience strongly suggests that a modular approach to large-scale change works best, each success increasing experience and creating a favourable climate for the next round. But avoid overcaution too. Although overambitious technological leaps invariably cause problems, projects spread over more than three years tend to get bogged down as the originators leave and 'ownership' dwindles. Timing is an aspect of change which often gets insufficient attention at the conceptual planning stage. Implementation means implementation of a defined project within budget and on time. A clear understanding of the time requirements, with milestones marked along the route, is an essential element in the equation.

Case study: Honeywell Information Systems

In 1985 the Scottish factory of Honeywell Information Systems, now absorbed into the French firm Bull, undertook to raise output of its DPS6 minicomputers from 600 to 1000, with 1500 in prospect for the following year. As well as increasing volume, Honeywell wanted to address other important business issues: improving responsiveness to the market on the outside, while holding or decreasing present inventory levels within the firm. In short, the brief was to turn it into a lean, flexible, world-class facility for building computers.

Feasibility studies quickly showed that Newhouse could benefit from advanced CIM applications, with extensive automation of handling, storage, assembly and test. But initial changes centred on much simpler techniques of JIT manufacture and quality improvement.

As a first stage, the company introduced '*kanban* squares' between each station on the DPS6 production line and gradually reduced the work in progress in each. In textbook fashion, as WIP shrank, it rapidly identified one after the other the constraints in the production system. At first the configuration of machines to match sales orders, carried out by staff technicians on a shift basis, was the limiting factor. When this was fixed, by re-engineering the task so that it could be done by hourly paid operators and taking on two extra people, the bottleneck shifted to shipping. When that was remedied, it moved to final inspection and then to the test area. But as each bottleneck was removed, the capacity of the line moved up: from 17 systems a week before the improvements to 40 after, with just an 8% manpower increase. Eventually the constraining factor moved off the production line and into the clerical, premanufacturing area: translating orders received into signals to the line.

The gains from the JIT techniques were remarkable (Figure 8.3):

- Lead times cut by 75%,
- Inventory turn up 50%,
- WIP reduced by 65%,
- Defects at test cut by 50% as the JIT feedback loops quickly relayed quality problems back to operators,
- More than doubled capacity,
- Unquantifiable benefits in making the plant easier to manage: a process hold-up was immediately visible to anyone walking around the shop.

By slashing lead times, these essentially low tech changes took over £2 million out of Honeywell's work in progress. This not only helped fund the subsequent phased four-year implementation of new technology (initially manufacturing cells with automated assembly and test), it also created the simple structures in which automation is easiest to apply and has most benefit.

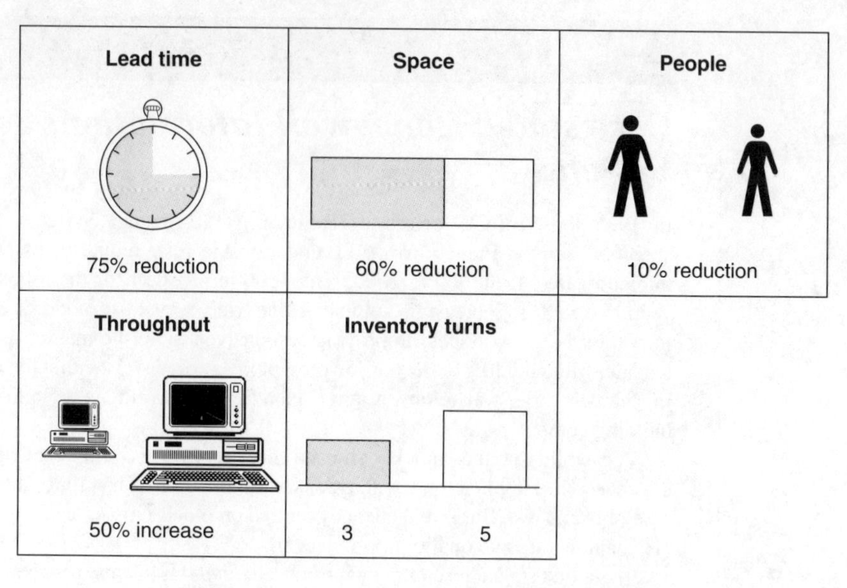

Figure 8.3 Honeywell: benefits from just-in-time.

It is important to stress again the distinction between the implementation of individual projects and the movement towards continuous change and improvement. The principles of project implementation are by now well tried and proven. Over the years project management teams have learned by experience how to meet production targets for major tasks, including management systems and software as well as manufacturing systems equipment. If they are built on the foundation of pre-automation product and process redesign, the actual implementation of advanced technology projects ought to hold few terrors for an experienced management.

But as these changes interact, becoming both effects and causes of further change, the wider implications for organization and management as a whole are much harder to predict. Drucker believes that the coming of the information-based organization is as epoch making as the earlier shift from owner to professional managership and the introduction of the modern decentralized command and control organization *à la* GM or GE (Drucker, 1986). Apart from the decline of the traditional middle manager and the rise of the 'soloist', he foresees the need for a new task force type of operations management, fluid organization 'beyond a matrix' and new disciplines of 'information responsibility'. This transformation awaits an Alfred Sloan or Georg Siemens to formulate it in definitive form. In the meantime, how do managers needing to take urgent practical action make sure they are facing in the right direction, with the right tools for the job?

Guidelines for manufacturing improvement

Upgrading manufacturing management

Through the whole change process, there are several guidelines to maintain consistency of goals. The first is the need to upgrade manufacturing management. For far too long, manufacturing managers have ranked near the bottom of the status heap, both in companies and in society as a whole. Perhaps significantly, the two exceptions to this generalization are West Germany and Japan.

Using advanced manufacturing technology, the emphasis shifts rapidly from managing physical to managing intellectual assets. One witness to this is the changing composition of the workforce: in Japanese FMS installations, Jaikumar (1986) found engineers outnumbering production workers three to one. This, he argues,

> 'signals a fundamental change in the environment of manufacturing. Flexible automation shifts the arena of competition from manufacturing to engineering, from running the plant to planning it. In the FMS environment, engineering innovation and engineering productivity hold the keys to success. Engineering now performs the critical line function. Manufacturing has become, by comparison, a staff or support function.'

In the same way, the operations focus also changes: from run time to set-up time, from design as an end in itself to design for manufacture, and so on.

There is thus urgent need in both low- and high-tech environments to develop teams of highly qualified managers who can unite business and engineering knowledge for continuous improvement. This is already the case in Japan, where FMS users have four times as many engineers and CNC trained personnel as their US counterparts (Caulkin, 1987). Note that they are engineers rather than IT specialists. An extremely important role falls to first line supervisors, on whom most of the day-to-day responsibility for timely production and high quality devolves.

The learning organization

There is no short cut to the development of the skills needed for JIT, total quality control, computer aids and intelligent automation. There is likewise no short cut to the physical implementation of complex CIM-type installations. The experience of achieving one phase of implementation is the necessary building block for the next.

This is why attempts to leap-frog straight to CIM have such a poor record. Note the painstaking approach of the Japanese in refining FMS, CIM factories in miniature, to a high level of reliability before moving to the next stage of automation. Machine tool maker Yamazaki Machinery Works is now on its third FMS facility, a colossus containing 88 machines and 32 robots. Following its successful flexible manufacturing factory (two FMS, giving huge savings in space, process time and manpower), Yamazaki is building its precociously named Factory of the 21st Century comprising eight FMS controlled by a central CAD/CAM centre 20km away via fibre optic lines.

In the case of equipment suppliers like Yamazaki such an approach is doubly effective: its process improvement is also its product R&D. As the performance of the advanced Japanese companies has shown, in information-intensive manufacturing this kind of intellectual asset is the key to competitive advantage. The new attitude to manufacturing is pithily summed up by the director of a western computer firm:

> 'Very often in the West, manufacturing is viewed as a pain in the rear, which you have to put up with to deliver the goods. We take the view that manufacturing can give us an edge in flexibility, responsiveness and service as well as serving as a test-bed for our own manufacturing products.'
>
> (Caulkin, 1987)

Time, speed, flexibility

In the past, the trade-off theory has ruled manufacturing. A manufacturer could aim at quality or low cost, volume or variety, and beyond a certain point gains in one could only be made at the expense of the other. But it is now clear that the trade-off is another piece of western manufacturing baggage which must be jettisoned. It is still true that a 'focused' factory which has clearly identified strategic aims and plays to its strengths will outperform the 90% which have no strategy. But now manufacturers have discovered that conflict between quality and efficiency is unnecessary, they are beginning to look at flexibility and fast response in the same light: not as an alternative but as an addition to manufacturing strengths.

The corollary, according to the international group of researchers at INSEAD, is that 'the next competitive battle will be waged over manufacturers' competence to overcome the age old trade-off between efficiency and flexibility' (De Meyer, 1987). Analysing the findings of surveys of manufacturing intentions conducted in the USA, Japan and Europe, they comment that while Europeans and Americans are battling to remedy quality and cost efficiency shortcomings compared with Far Eastern competitors, the Japanese have moved on. They are now bent on developing a new competitive edge by combining low cost, high quality manufacture with flexibility and speed of response. If the decade 1975–85 saw the demise of the quality/efficiency trade-off, 'the decade 1985–95 has the potential of becoming the era where manufacturers will discover that flexibility in all its aspects is not necessarily contradictory with the pursuit of cost efficiency'. If this is the case, says the INSEAD group, 'the Japanese competitors seem to be further down the road in making this "cost-efficient flexibility" into a reality'.

For most western manufacturers 'cost efficient flexibility' actually means reducing chaotic and overextended product ranges in order to focus strengths and make flexibility manageable. Flexibility to meet changing market needs is affordable and adds value; flexibility to cope with ramshackle products and processes adds cost. To meet a focused Japanese challenge, SKF could no longer make its whole range of bearings in every factory; it had to concentrate each one on two or three major lines. In their plant and equipment choices, companies should avoid overflexibility as a major cause of expensive complexity. If an FMS is repetitively turning out the same parts, there is no point in having it. For focused flexibility keep solutions simple and avoid foreclosing technological options: another reason for a modular, resilient approach.

Case study: John Deere

The giant farm equipment manufacturer (John Deere) broke fresh ground a decade ago with factory automation that was then regarded as the model for all others to emulate. FMS technology worth $1.5 billion was designed to provide a choice of 5000 process changes on ten basic tractor models.

As Deere had to weather a depressed farm equipment market and then a strike, it became apparent that the company had invested too much in its state-of-the-art FMS without regard to the process being automated…

Billion dollar losses were followed by rationalization of the production process, and today automated manufacturing at Deere is a much more organized, simplified affair. John Lardner, a company vice-president, said: 'The FMS was a retrofit to a production design problem that shouldn't have existed in the first place' (Warner, 1987).

Organizational skills

Organizational improvement is a better and cheaper way of managing complexity than technology. Instead of installing MRP, try to do without it by reorganizing factory layouts to accommodate group technology and JIT. Managers brought up in traditional structures and lacking role models will find problems

in these areas easier to define than to solve. It is one thing to say, correctly, that functional divisions must be broken down in the cause of integrated manufacturing, quite another to anticipate the long term effects on career patterns, remuneration and the like. It seems certain that the resulting structures will be more fluid and less permanent than in the past, with strong emphasis on task-based teams. Design for information will be a key structural principle, based on just-in-time information flows. JIT already contains a bias towards simplification and selection of information on a 'need to know' basis. The job of making specialists more line oriented will entail much more job rotation than at present. As manufacturing flexibility becomes routine, organizational innovation will become the key to the next struggle for competitive advantage, greatly increasing the requirement for managers with organizational skills.

References

Caulkin S. (1987). ICL's Lazarus Act, *Management Today*. January

De Meyer A. *et al.* (1987). *Flexibility: the next competitive battle*. INSEAD Working Paper 86/31

Drucker P. (1986). *The Frontiers of Management*. Heinemann

Ernst & Young (1992). *Process Innovation – The UK View*

Haughton E. (1992). Moving the corporate goalposts. *Computer Weekly*, July 30

Ingersoll Engineers (1985). *Integrated Manufacture*. IFS, Springer Verlag

Jaikumar R. (1986). Post-industrial manufacturing. *Harvard Business Review*, November–December

Skinner W. (1988). What matters to manufacturing. *Harvard Business Review,* January–February

Warner T. N. (1987). Information technology as a competitive burden. *Sloan Management Review,* Fall

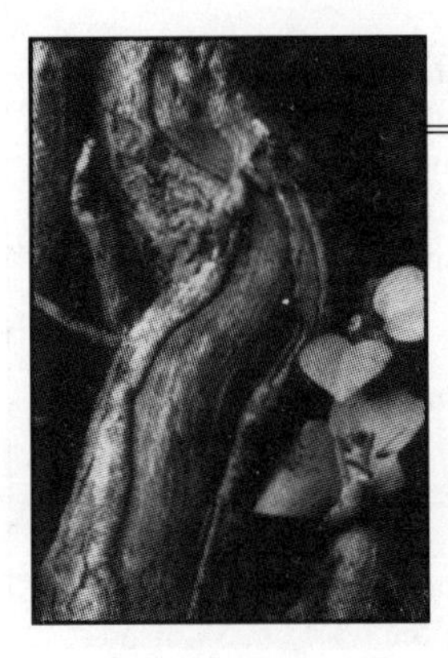

9

World class manufacturing

This is the ultimate goal for any right-minded manufacturing company – to aim for excellence. How does one define 'world class manufacturing'? One source states

> 'it essentially entails having the right production capability to make money from totally satisfying the customer, with high-quality services and products at the right price, delivered at the right time. It means operating at standards equal to the best in the world, but it is not relevant only to companies that export – it is just as relevant to companies that face overseas competition, and that's just about everyone.'
> (DTI, 1991)

Case study: Dowty Aerospace

In the last ten years, Dowty Aerospace has become the largest and most profitable division of the entire Dowty Group.

From this position of strength Dowty Aerospace undertook in 1990 a two year corporate restructuring in order to capture a greater share of world markets, particularly North America. To this end it opened a new £45 million manufacturing plant in Montreal in 1991.

continues

continued

Further investment in the latest machining techniques took place. For example, the production of high-value, complex and accurate aerospace parts was transferred to Hitachi Seiki HG series machining centres, equipped for unmanned running. These high-specification machines, operating as machining cells and capable of expansion and linking up are considered to be essential technology in the competitive world of aerospace.

Dowty completely revamped its old corporate structure. Out went the old geographic structure, where separate UK, North American and worldwide operations each contained a mix of activities. In came five reconstituted multinational businesses, each concentrating on a specific product or service area.

This regeneration of corporate identity, plus strategic investment in the latest technology, has taken Dowty Aerospace into the world class manufacturing league.

Aiming for excellence requires a three phase strategy. First, there is the preparation, talked about in earlier chapters of this book: understanding the competitive global climate and your own company's market position, analysing specific corporate needs, setting up a business strategy, developing a TQM philosophy, anticipating risks in the operational functions and preparing the workforce. Second comes the practical application of manufacturing techniques, implementation of new technology, upgrading existing technology, reorganizing methods and practices and training or retraining the workforce. Some of the practical applications are dealt with in this chapter; the human factor is dealt with in a later chapter. The third phase is managing the change by developing all the strands of the manufacturing process in conjunction with a business strategy – never losing sight of the aims of the company or the needs of the workforce – keeping control as you take the company into the world class category.

Yet more buzzwords

Before getting down to the nitty-gritty it is worth getting some of the buzzwords of the 1990s out of the way, so as to avoid confusion. Aiming for world class manufacturing requires a company to follow a fairly logical approach based on the requirements outlined above. Although the terms used to describe this approach can vary, depending on which school of thought one follows,

World class manufacturer checklist

According to the DTI, a world class manufacturer would score 10 out of 10 by being able to answer 'yes' to the questions in on the checklist below.

(1) Is your plant clean, tidy and uncluttered?
(2) Is your production facility or process completely dependable and reliable?
(3) Is your design, production and other documentation clear, up to date and always used correctly?
(4) Do you attach sufficient importance to developing your product and process engineering?
(5) Is your workforce totally flexible?
(6) Are you always achieving the shortest possible throughput times?
(7) Are you committed to total quality, with a plan for continuous improvement on all performance measurements, including new products?
(8) Are you actively committed to training, retraining and competence throughout the whole firm, including management?
(9) To what extent do you consider the shop floor as a source of ideas?
(10) Do you accept that continuous change will be the future pattern of life in manufacturing and act accordingly?

they all lead to the same goal – make no mistake about that. As one chief executive of a French electronics company said,

> 'We wasted a lot of time in the boardroom fighting about which methodology to use in order to achieve our aims, when what we should have been doing is absorbing elements from all the methodologies and working to a plan that *we* created, that suited *us*. We got there in the end, but I can't help feeling that it took us longer because first we had to decide on a method and then we had to try and shoehorn our company into it.'

(1) CIM, CIB or HCIM?

Computer integrated manufacturing (CIM) and computer integrated business (CIB) were described is Chapter 2 as, basically, the integration of all processes within the factory (CIM) and the integration of all other businesses activities with the processes in the factory (CIB). Human and computer integrated manufacturing (HCIM) is an almost embarrassing afterthought. Should we really need to be told in the 1990s that we should integrate the organization, human resources and technology in order to produce a dynamic manufacturing environment? This means not just the interface between people and computers but also people with other people in the organization. Are we, 40 years after the computer was born, only just considering the contribution that people make to the manufacturing process? Shame on us in the West if we are. However, it is a term bandied about in this decade.

(2) Balanced manufacturing response

Balanced manufacturing response is an umbrella term for developing a strategy along the HCIM lines. It means not letting any one element of a company's operation, such as technology, dominate the others, such as people and organization.

(3) Lean production

A concept launched upon the world by the Massachusetts Institute of Technology, lean production basically promotes less of everything – less human effort in the factory, less manufacturing space, less investment in tools, less engineering hours to develop new products, and in less time. It looks to reduce changeover times from hours to minutes, to avoid the need for expensive systems to track materials, and to arrange manufacturing into cells which can achieve a natural rather than an imposed momentum. Lean production means utilizing highly flexible machinery run by teams of multiskilled workers who have a commitment to total quality. This climate of high responsiveness will actually mean that a company produces lower volumes of goods but in a greater variety than before.

(4) Simultaneous engineering

Another integrated approach, simultaneous engineering is applied to new product development. To many companies it would demand a radical change in thinking since it requires that research, design, development, manufacturing, purchasing, supply, marketing, service and maintenance all work together to develop a product from beginning to end, rather than coming into the development process one after the other.

The Japanese manufacturers have used this technique successfully for many years and the system appears to cut new product times by 75%, cut assembly times by 50%, reduce engineering changes and piece parts and build quality control in throughout the process.

(5) The rational factory

In the rational factory facilities, layout and operating principles are configured to provide optimum response to support the needs of the business and the correct requirements of the marketplace (DTI/PA Consulting Group 1989). The key constituents are: effective sourcing; effective processes; flexible factory, flexible automation; environmentally positive factory; dependable factory and devolved organization and layout.

Methodology madness

(An extract from *Peopleware* by DeMarco T. & Lister T., Dorset House Publishing.)

Of course, if your people aren't smart enough to think their way through their work, the work will fail. No Methodology will help. Worse still, Methodologies can do grievous damage to efforts in which the people are fully competent. They do this by trying to force the work into a fixed mold that guarantees:

- A morass of paperwork
- A paucity of methods
- An absence of responsibility, and
- A general loss of motivation.

The following paragraphs comment on each of these effects.

Paperwork

The Methodologies themselves are huge and getting huger (they have to grow to add the 'features' required by each new kind of situation). It's not at all unusual for a Methodology to use up a linear foot or more of shelf space. Worse, they encourage people to build documents rather than do work. The documentary obsession of such Methodologies seems to have resulted from paranoid defensive thinking along these lines: 'The last project generated a ton of paper and it was still a disaster, so this project will have to generate two tons.' The technological sectors of our economy have now been through a decade-long flirtation with the idea that more and more paperwork will solve its problems. Perhaps it's time to introduce this contrary and heretical notion: *voluminous documentation is part of the problem, not part of the solution.*

Methods

The centerpiece of most Methodologies is the concept of standardized methods. If there were a thousand different but equally good ways to go about the work, it might make some sense to choose one and standardize upon it. But in our state of technological infancy, there are very few competing methods for most of the work we do. When there are genuine alternatives, people have to know about it and master them all. To standardize on one is to exclude the others. It boils down to the view than knowledge is so valuable, we must use it sparingly.

continues

continued

Responsibility

If something goes wrong on a Methodology effort, the fault is with the Methodology, not the people. (The Methodology, after all, made all of the decisions.) Working in such an environment is virtually responsibility-free. People want to accept responsibility, but they won't unless given acceptable degrees of freedom to control their own success.

Motivation

The message in the decision to impose a Methodology is apparent to all. Nothing could be more demotivating than the knowledge that management thinks its workers incompetent.

Where to start?

From a writer's point of view it is difficult at this point to know where to start since, effectively, the two major issues – developing the right sort of factory (rational or lean) and integration, should go hand in hand. Theoretically, the type of factory and how its processes, people and machines meld together to achieve the aims of the corporate strategy should be considered in the initial planning stage. However, very few manufacturers are in the fortunate position of starting with a clean sheet. Most have to re-engineer what they have in order to achieve optimum results (see Chapter 10). Therefore, for the purposes of this book, we will address the problem of the organization and equipment within the factory first, since it lies at the core of any competitive manufacturing strategy.

Sourcing

Lean production, remember, means stripping out of a factory that which is not cost- or time-effective and, in order to do this, one must first determine which is the best source for products, components or manufacturing processes. It may be that outsourcing would provide a manufacturer with a cheaper, faster and better quality component than by doing it in-house. However, it could also be that certain suppliers have a monopoly on a particular component and they have artificially raised its price. To invest in the processes to produce that component in-house, could eventually pay considerable dividends.

Therefore, an extensive analysis has to be undertaken to determine the most effective method of sourcing. The manufacturer with a significant export market might consider, for example, sourcing an element of the product in the country where it is eventually sold, to minimize the effect of currency fluctuations. When assessing external suppliers the prime concerns should be the capacity, flexibility, skill, reliability, cost and quality available. When assessing internal sourcing the questions are will it save money, do we have the skills and capacity, how much will we have to invest to develop the skills and capacity and what will be our return on that investment, and what impact will internal sourcing have on the rest of the organization?

Processes

The traditional types of manufacturing processes are line process, continuous process, batch, jobbing and project. But technology has allowed the types of processes to become modified or adapted so that the categories are not so clear cut anymore. However, let us briefly examine the traditional choices.

Continuous process

A the name implies, the value of this system is that it should run continuously with minimum shutdowns, therefore it is the choice of manufacturing companies that have high volume demand, such as liquid or gaseous fuels, or food and drink. In continuous process there is very little need for manual labour – the materials are automatically transferred from one part of the process to another. Any human involvement is usually supervisory, quality checking or maintenance.

Line process

This process is usually dedicated to making a small but standard range of products, for example domestic appliances, which are produced using the same sequence of operations. High volume but not very flexible.

Batch

Using batch methods is best suited to companies that produce similar items in varying qualities. The process involves setting up 'islands' of manufacturing which contribute to the manufacture of the whole. Machinery and tools may have to be altered for each 'batch'. An example of this would be a print factory where each product is similar but requires new plates, different inks, different paper and different methods of binding and packing and where the order could be high or low volume.

Jobbing

This is an expensive and time-consuming area of manufacturing since it involves the on/off order requirement for a product that will probably never, or rarely be needed again. Examples vary from the manufacturing of a limited edition of a china figurine to the manufacture of a purpose-built tool.

Project

A process required for one-off, large-scale projects, for example, constructing a communications satellite over several years. Different sections, components and materials almost certainly have to be made in various places and brought together to make the whole. This whole, scattered process of manufacturing, the costs involved, and the deadlines to be met, have to be managed centrally as a project.

But, as mentioned earlier, technology has affected the traditional choice of processes available to the manufacturer. Process substitution is a large area of research in Japan, where companies are busy identifying the nature of the process that they need and then designing new equipment, or modifying an existing design to create a hybrid process.

The more common hybrid processes are outlined below.

Numerical control (NC) machines

This is a process whereby a machine performs a task to a high degree of accuracy by following a set of coded data, which can be input by card, tape, computer disk or manual. Traditionally NC machines are used in heavy industry for metalworking processes. Most modern NC machines are controlled by microcomputer and were the first examples of computer aided manufacturing (CAM).

Machining centres

These combine several NC operations into one unit. Typically, a machining centre would stand at a single location (by the side of a product which may require various machining operations). A carousel holding up to 100 tools rotates according to the instruction data received, enabling it to perform many machining tasks (usually in a set order) without disruption.

Flexible manufacturing systems

These systems combine NC, CNC and/or robots, with automated materials transport and handling systems, controlled by a central computer. The computer can control the actions of individual machines and regulate the flow of production from delivery of source materials to delivery of end product. Flexible manufacturing systems (FMS) are the core of computer integrated manufacturing (CIM) – CIM effectively includes the strategy, design and management of FMS.

The DTI has suggested that the following technologies will prove important in the 1990s:

- Near net shape processes (moulding, powder metallurgy, precision casting/forging, superplastic forming, flow forming, and so on),
- Sensors,
- Application specific integrated circuits,
- Composite materials,
- Expert systems,
- Simulation,
- Integrated materials handling.

Selecting a process

To select the right process or combinations of processes requires an understanding of which will best meet the specific company's needs of cost, service and quality. Does the process allow for a variety of components to be processed without disruption or allow several operations to be combined? Could any manual decision-making processes be replaced by an expert system? The manufacturer needs to assess potential costs at the same time as assessing potential risks. For example, would any high-tech process be flexible enough to cope with future 'unknowns' – such as change in consumer demand and changes in labour and skill availability? Also to be considered must be the impact that any new factory process has on the internal and external environment. Will the workers be bored, get dirty or be put at some sort of health and safety risk? Will the outside environment be affected? What is the legislation regarding such matters and how much will it cost to meet the requirements?

Factory operation and automation

Experts suggest that the factory of the future must avoid investment in high-inertia capital assets. The depreciation adds considerably to the overheads, apart from other disadvantages. The message is: low inertia to create maximum flexibility. The low-inertia factory uses individual plant items of high efficiency, but relatively low unit capacity – which is adjustable as the market demands. Low-inertia factories should also be able to retool with minimum cost and disruption and reconfigure the general layout fairly easily.

In creating a flexible factory the manufacturer needs to ask whether the existing flow of material is logical and orderly and, if it were changed, is there space to do so or to extend with very little disruption. Various factory layouts can be tested using some form of computer simulation to see whether various ideas are workable.

Flexible manufacturing also requires a flexible workforce who are trained to be willing and able to move from process to process.

Case study: Yamazaki Mazak

This UK-based Japanese machine tool company has achieved a high quality, low cost environment within its factory while also achieving maximum flexibility. The company makes 55 different products as opposed to the traditional 15 offered by competitors. These are produced with a combination of a multiskilled workforce, advanced manufacturing technology and flexible working practices. In addition, the factory operates with a manufacturing lead time of two months, compared with the industry average of six months.

Integration

Integration starts with analysis of functions, people, machines, systems and the current and future needs that will drive the manufacturing business. Then the 'communication' links or interfaces between each of those are determined. Changes are planned for at this stage: for example, the change from a manual system to a computer-based system to provide a higher speed of response.

In the experience of Derek Nunney, author of *Integrated Manufacturing* (DTI, 1992) the following areas are those which are commonly in need of significant improvement in response time:

- Product definition/production planning – this is especially in relation to early warning of changed or new bills of material (BOM).

- Design/production engineering – it should be an aim to use common design data without the need for recalculation or interpretation, for example for method study, tool design and so on. There should be early access to changed or new designs.
- Design/production planning – this relates to plant requirements and production capacity.
- Planning/manufacturing – this refers to the need to manage and coordinate the information crossing this principal interface, so that a significant change in any one function can be assessed in terms of its effect on the whole business.

Simulations are useful

Because integration is a complicated business, computer modelling is useful in the planning stage to create various scenarios of man/machine interfaces within the organization and to build in an element of 'what if' different factors change due to external pressures.

Most integration exercises, because of their size, are usually undertaken in phases, along a planned migration route – new manufacturing facilities being developed in tandem with systems development. Moving towards the integrated business in phases obviously means that technology will be introduced over a lengthy period of time, which is why most companies are now trying to use products that conform to Open Systems standards – to allow ease of data communication between systems.

Information systems must be planned and managed and assembly must be integrated with component production to achieve business goals. 'Islands of automation', where they exist, can be integrated successfully, as can transportation and communications systems. Ambitious integration plans include suppliers and the logistics operation too – bringing the whole together under a computerized factory control system which manages the changes in quantities, timings, variations, while monitoring production achievement and directing labour resources.

Case study: ASEA Brown Boveri Traction

Swedish company Asea Brown has utilized an integrated engineering information system which has substantially cut down the time taken to develop a new product and bring it to the market.

continues

continued

The system stores the complete description of each new product as it is designed. The description includes the bill of materials, geometry of mechanical parts, the layout and logic of electronic systems, strength calculations of the structure, process plans and much more. Each designer, whether in the head office or in another country, can access this information and input his or her own data. The system is linked into the local and international communications network and has built-in change control and access security facilities.

MRP vs JIT vs OPT

Because there are so many variables, allocating plant resources to maximize productivity or profits is a huge mathematical task. In a traditional factory where hundreds of workers make a variety of products on a range of different machines, a part could follow an almost unlimited number of routings. In most factories, the reality is so far from the mathematical optimum that

> 'even a clearly suboptimal solution may offer substantial improvement. The need is for a relatively simple method of planning and control that can cut down on long waiting times and eliminate most of the inventory costs.'
>
> (Gunn, 1982)

Production control is a task in which computers evidently have a part to play. But unlike, say, CAD, which depends for its effectiveness on computational power and clever programming, the issue in resource planning and control is a systems one: what model to adopt for the functioning of the plant as a whole? Compare, for instance, the characteristics of the three most commonly used systems: MRP II, JIT and OPT.

MRP, first developed as a manual technique (materials requirements planning) in the 1950s and elaborated by IBM into the computerized manufacturing resource planning tool a decade later (designated the MRP II), is a sophisticated form of back-scheduling: planning the input of materials, labour, machine time and other resources into the production process by extrapolating backwards from a delivery date for the finished product. As a database of parts and their relationships, finished goods, WIP and lead times, MRP aims to manage parts inventory more effectively. By predicting when and how many of each part is needed, scheduling aids manufacturing just-in-time for use.

MRP functions by 'exploding' work into its constituent parts and 'pushing' them through the stages of manufacture to meet the master production schedule (MPS), based on orders and forecasts. A simple MRP system comprises an MPS, bills of material (defining all the parts which make up a finished product), inventory records and process plans to set lead times. A more complex package may contain up to 20 software modules, including production scheduling and links to financial accounting. To work it must be based on highly accurate data covering orders, due dates, inventory levels, capacities and lead times. It also required fast feedback of significant changes at subsequent stages of the planning and manufacturing processes. Hence the 'closed loop' denomination, akin to that used for process control in process industries.

Computer makers have pushed complex closed loop MRP as the key to manufacturing control; they want to sell more computers and planning software to run on them. Of UK manufacturing companies, 40%, especially larger ones, are said to use MRP. Its effectiveness has been most clearly demonstrated where pre-automation improvements in physical layout and organization have also taken place. The Cranfield study shows useful gains during the 1980s in lead times for component parts manufacture. But is is less clear that benefits at this level are carried through into the business performance of the organization as a whole. Typical problems are as follows.

- *Inaccurate master plans*. The MPS generally relies on forecasts. But the more it is based on forecast, the less it will match actual demands. First result: what is pushed into the system is not equal to what is being pulled out. Second result: either higher WIP and longer lead times, or shortages. Rerunning the MRP system is a complex and time-consuming job, often a matter of a weekend rather than overnight, and it can only be made to respond to the hour-by-hour changes of responsive manufacturing by huge expenditure on hardware and communications. Even them, the results are likely to be too complex for easy application.
- *Inaccurate inventory records*. Maintaining highly accurate records is difficult, expensive and adds cost rather than value. It requires heavy investment in stores and movement control, terminals, bar code readers and so on. The only way to outflank the system's need for accuracy is to abandon closed loop control.
- *Inaccurate lead times*. Actual lead times vary enormously according to a range of factors: batch size, scrap rates, machine utilization, motivation and so on. They are therefore often set longer than the optimum to ensure that the MRP system is provided with potentially accurate data, but at the cost of some of the benefits in WIP and shorter lead times that the system is supposed to achieve.
- *Shortages*. The result of these problems is often shortages, and the need for 'hot lists' and expediting systems to manage them. These not only further worsen the accuracy of the official system, they tend to take over control from it, discrediting it in the process. Actually, hot lists and shortages are the cause, not the symptom of the problem: they represent actual customer demand, or what the factory should have produced if the MRP scheduling had been correct.

- *Quality costs*. MRP has no way of separately monitoring quality costs such as variable yields, scrap or rework. As for lead times, a pessimistic assumption is typically built into the calculations, which gives operators little incentive to do better. MRP as a tool does nothing to manage or improve quality, although it is often touted as the total answer to the manufacturing problem.

Unlike MRP's 'plan push', JIT-oriented systems operate in a 'pull' mode. Actual orders trigger a finishing process from which the demand signal cascades back through preceding processes until it reaches suppliers. Materials move continuously through the factory, pulled onwards by the next stage of manufacture. Since information loops are short, control is decentralized, although some kind of overall planning is necessary. Computer support is correspondingly less, typically taking the form of networks of PCs rather than a central mainframe. JIT is generally considered most effective in repetitive, high volume manufacture. There are three reasons why it sometimes works less well than expected.

(1) Long procurement lead times. Western JIT converts are sometimes frustrated to find they do not have the clout to persuade suppliers (particularly large companies) to deliver little and often. The company must therefore use a forecasting mechanism to supplement the demand pull signals in order to determine supplier schedules.
(2) Demand variability. Without forewarning of demand peaks and troughs, individual processes must be sized to meet peak demand. The Japanese simply invest in the extra capacity. To western minds this is not cost-effective, and in most cases capacity is set to average demand. This means that some kind of forecasting and inventory build-up is needed to deal with the peaks, swamping capacity, interfering with the demand pull signals and destroying much of the point of the JIT techniques used.
(3) Timidity. JIT does not work without whole-hearted integration. Since levels of contingency are low, commitment to quality and planned maintenance, for instance, are essential.

After a period of head-on rivalry, MRP and JIT are now seen as complementary. MRP is a sophisticated planning tool which falls down when it is asked to control manufacturing with inadequate information. It also lacks any built-in quality reinforcement. JIT is a simple and effective control mechanism, with a strong bias towards quality, which works less well in situations calling for complex planning. Recent thinking attempts to combine the strengths of the two, using MRP for planning and JIT for execution control. This implies on the one hand abandoning the closed loop and explicitly decoupling planning from control; and on the other hand, introducing some simple shop floor level techniques to modify the 'demand pull' signals to take account of major variations in demand or supply. Some physical rearrangement of processes to smooth material flow and establish cellular manufacturing techniques is also usually necessary.

Optimized production technology (OPT), unlike MRP and JIT, is a proprietary system developed in the 1970s in the USA and available in the form of a software package and associated control techniques licensed from a single supplier, OPT has points in common with both JIT and MRP, and users claim impressive results. OPT is a scheduling system which focuses on the progressive removal of bottlenecks in production in order to improve 'throughput' (cash generated by sales) while cutting inventory and operating expenses. Like JIT, it concentrates on issues like quality, machine set-up, lot sizes and lead times, varying the last two according to calculations of finite capacity. At the planning level, OPT has certain similarities with MRP, which it also resembles in demanding a detailed database of product and machine information. The complex software can only be run on minicomputers or mainframes. OPT is expensive, putting the software if not the associated consultancy out of reach of small firms, and the chequered history of its founding firm may be considered a drawback by some.

What goes wrong

- *Technology addiction.* Classic manifestations are FMS which are not flexible, MRP systems requiring extensive tailoring, computers where paper and pencil will do. Overcomplicated technological solutions are the result of paying too much attention to persuasive vendors, either external or internal, and not enough to business needs. They are the symptoms of computer driven rather than business driven planning, technology-push rather than problem-pull solutions.
- *Too bold a technological leap.* Moving directly to automation is a high risk approach. Given the lag in manufacturing performance, it is one which European and US firms find tempting.
- *Half-heartedness.* This is frequently the consequence of mediocre results from previous piecemeal investment in new technology, itself stemming from the lack of a proper plan, a lukewarm top management and the absence of a champion. Any one of these three faults will kill a project stone dead.
- *Inaccurate and inappropriate data.* Elementary and surprisingly often overlooked; frequently the result of clogged and inefficient systems. That is why manufacturing plants became uncompetitive in the first place. Is all that inventory recording actually needed? Do the bills of material match how the products are actually built? Do the management accounts tell you what you need to know to run the business?
- *Underestimating software difficulties.* In-house software projects notoriously obey the 2-2-$^1/_2$ principle: they take twice as long, are twice as difficult and provide half the benefits expected. Prefer standard packages until you really know what you are doing.

continues

continued

- *Failure to attend to detail.* Although the technology is complicated, manufacturing efficiency is fundamentally about doing simple things superlatively well. That means unswerving attention to detail at the implementation level, to complement wide strategic vision at the top.
- *Inflexible organization.* More responsive manufacturing demands the breakdown of traditional territorial jurisdictions. Without the willingness to experiment with new organizational forms, many of the benefits of new technology are lost.
- *Inadequate communications,* invariably accompanied by a failure to involve the workforce. Integration means using all the resources of the manufacturing business. Many companies fall down by privileging the technology or the integration, and forgetting the people.
- *Under-resourced implementation,* generally the result of planning deficiencies. The faulty understanding can often be traced back to a failure to define proper responsibilities and lack of commitment at top management level.
- *Insufficient training.* The important of training and retraining, from the bottom to the very top, is fundamental. Training and implementation go together. Together, the 'people' factors (training, communications, involvement) are estimated to be the most frequent single cause of failure in AMT programmes.

Logistics

The final permanent principle in planning and implementing change is the need constantly to review the logistical base of the organization, both inside and out. GM's realignment of purchasing as 'supplier development' has already been noted earlier. This is more than a matter of semantics. Manufacturing no longer starts and finishes within the factory walls. Under the pressures of Japanese competition, US auto companies are moving from a traditional strategy of corporate independence towards a collective strategy in which stable groups of customers and suppliers are recognized as interdependent: the health of Ford, Chrysler and GM at the top of the supply chain is largely dependent on the health of their supplier community, and vice versa.

Within the supplier/manufacturer continuum, the areas of responsibility are changing. In the car industry, the decision of the auto makers to focus their activities to reduce response time is restructuring the industrial base. Between the car makers and the providers of commodity parts there is a new layer of 'first tier' suppliers of sophisticated subassemblies such as integrated windows, seal and winding mechanisms or complete doors. These companies no longer compete with each other solely on cost. To survive, they must themselves develop cost saving and quality boosting innovations for products and processes. Both by early involvement in the design process and by electronic data interchange, they are increasingly integrated with their customers. Novel in the West, this kind of relationship is common in Japan, where groups of related companies are often linked by shareholding. It is likely to increase, with important implications for the management of quality, design, supply and distribution, in industries other than cars. Aerospace manufacturers, for instance, are subcontracting major elements of commercial airframes to single suppliers. Integrated logistics is the competitive battleground of the 1990s (Kearney, 1988).

Golden rules of implementation

These are the constants that senior managers must bear in mind as they implement programmes which lead towards more responsive manufacturing and set the ball of continuous improvement rolling. But the most important success factor in any advanced manufacturing project predates the implementation stage: it is getting the strategy right in the first place. The rule of thumb is that the greater the attention spent getting the strategy correct, the less problematic the process of putting it in place. Too little initial feasibility and planning preparation builds up to overwork in the implementation stage, late completion and, even worse, only partial achievement of the objectives. See Figure 9.1.

The key is to look closely at the systems level to work out the detail needed to support it. For example, 3% faults on a part may be manageable (although costly) when there are humans around to do the rework. In a highly automated plant driven by JIT, that failure rate would cause havoc. Or take the case of a new automated guided vehicle (AGV)-fed manufacturing system. In the past, defective pallets were no problem: they could be fixed in a minute by a man with a hammer. In the automated system, nonconforming pallets jam the mechanized store and fail to pass monitoring gauges on the shop floor. Some manufacturing radicals reject computer simulation techniques (Schonberger, 1986) for modelling new factory methods because they fail to pick up these detail implications and encourage overcomplicated solutions. They are useful, however, in investigating complex systems, and consultants claim large benefits for simulation in planning the phasing of capital expenditure, developing unfamiliar manufacturing systems with complicated knock-on effects, identifying layout design faults on the computer screen rather than the factory floor and cutting out 'just-in-case' equipment purchases. Savings made at the concept planning stage or in rapid implementation later usually pay for the 1.5% of proposed investment which simulation typically costs.

Having planned carefully – Fujitsu Fanuc took 100,000 man hours to plan one new plant – you must implement aggressively. This does not mean doing everything at once. It does mean recognizing the danger that momentum can

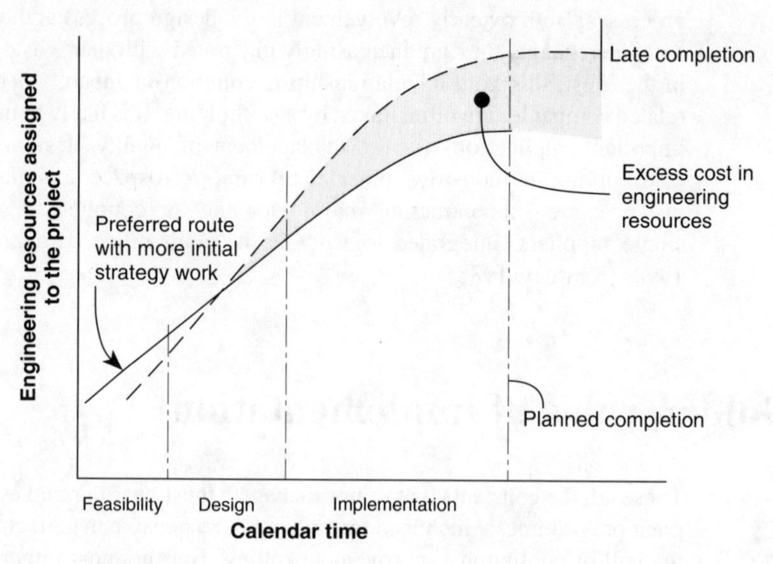

Figure 9.1 Cost effectiveness of upfront engineering effort.

slacken without effort to maintain it. This effort is beyond the capacity of people who are already fully occupied with day-to-day departmental duties. The only way to achieve the radical changes needed is by creating a full-time project team with authority to overrule the 'custom and practice' which can easily smother new projects. In such programmes, the Pareto principle almost always applies. The project could reap 80% of the benefits for 20% of the cost in the initial stages. But most of the risk occurs in the final phase, with the installation of costly new machinery. There is in the end no substitute for careful planning, both for technology and for the human element, which forms the subject of the next chapter.

References

DeMarco T. and Lister T. (1987). *Peopleware*. Dorset House Publishing

DTI (1991). *Aiming for world class manufacturing*. Department of Trade and Industry

DTI/PA Consulting Group (1989). *Manufacturing into the late 1990s*. Department of Trade and Industry

Gunn T. G. (1982). The mechanization of design and manufacture. *Scientific American*, September

Kearney A. T. (1988). *Logistics Productivity: the competitive edge in Europe*

Nunney D. (1992). *Integrated Manufacturing*. Department of Trade and Industry

Schonberger R. (1986). Simple systems, simple models. *World Class Manufacturing*. Free Press

10

Adding the numbers

Accounting on trial

The paradox of IT investment

For a long time, there has been much confusion among accountants and businessmen about the seeming paradoxes of IT in manufacturing. Why do some applications pay off brilliantly in financial terms while others are a dismal failure? Why, with all the much touted benefits of computers, robots and automation, is 'the factory of the future' apparently so hard to justify as an investment and so slow in arriving? Drucker (1987) ruminates on the question why, although the 'pay-off from automation is indeed fast and high', it is so difficult to convince top management to invest in it. Skinner (1985) puzzles over 'the ironic situation of industry loaded with unused technology yet in trouble in terms of competition, costs, flexibility for volume and product change, return on investment', and so on. What is going on?

One answer is that too many companies have superimposed new technology on ancient organizations, thereby automating problems, not solutions, and reporting disastrous results. Another, linked to the first, is that businesses are being driven in the wrong direction by their financial measurements. The convention is that numbers are simply a measurement of something done, a neutral tool. The fact is that the way the measurement is taken unavoidably influences the present direction of travel. To put it crudely, as soon as managers know what figures are being used to monitor their performance, they start massaging them.

Themselves an unreconstructed part of the old manufacturing, the financial criteria are actually part of the problem. Not surprisingly, they turn out to be a treacherous guide to the solution. The new manufacturing philosophy demands radical change in accounting practice, but more importantly in its whole orientation. As with the other functions, it requires an integrated partnership in which accounting follows, rather than leads (as it still does in too many cases), the manufacturing and business strategy; in which it feeds into the self-reinforcing cycle of continuous improvement rather than the vicious circle of inappropriate controls and diminishing returns. In short, the accountant must become an active partner in technological change.

The new need

To put this in perspective, accounting adds no value to the product. But it has enormous influence (particularly in the West) over the way the business is run. Appropriate accounting provides an information framework, perhaps the most important one, for tactical and strategic control. Inappropriate accounting can disguise a company's competitive situation, block needed investment and encourage lesser priorities. At the primary level, the accounting function must support the overall competitive strategy. That means recognizing the implications of the spread of new technology outlined in Chapter 1: issues such as the rise of the information economy; the place of information systems in the firm; the use of product and process technology as a competitive weapon; the globalization of markets; and the demise of the traditional manufacturing trade-offs.

In other words, it must recognize that the age dominated by short-term financial constraints and decision rules is dead. Suppose the rules demand a three year payback for an investment, but the opposition is investing on a five or eight year view. Or suppose an investment project in itself (a company-wide telecommunications system, for instance) has a low rate of return but forms part of an overall strategic advance. Does the company invest or not? Even more fundamentally, accounting must come to terms with the fact that the modern manufacturing business is no longer cost driven. Costs are i mportant. But they are determined by pre-planned levels of quality, flexibility and speed of response, not the other way about. In any case these are no longer manufacturing alternatives, either to each other or to low cost. In the new philosophy, they are complementary parts of high quality, low cost flexible manufacture. Not only are they not contradictory: high quality can be least cost, and vice versa. Accounting must reflect this fact.

The old accountancy

Three functions of accounting

Accountancy is a good example of the inability of IT in itself to provide better information. Accountants were the first corporate users of computers for routine functional purposes. But they have been extremely slow to use their number crunching power creatively, that is, to link it with production to generate more useful information about manufacturing performance. In general, accountants contented themselves with automating what they were doing already. The result: the same inappropriate information, but faster and in overwhelming detail.

Most accounting systems, like most factories, developed unplanned. Accounting has three functions. The first is to provide statutory information for audit. The second is to provide management information/accounts for day-to-day operations. The third is to provide longer-term financial analysis, including investment appraisal, for the future. Each of these has a different timeframe and a different data requirement, and no single accounting system is ideal for all three. Unfortunately, in many cases the identity of the systems gets blurred, the original purpose for collecting the figures is lost, everyone has to report everything and essential indicators are buried in a mass of irrelevant detail. Despite all the figures, the plant is still out of control, and more accountants are taken on to master it.

A caricature? When consultants recently investigated a UK electronics manufacturer, they found that of the 140 people in the accounting function, only six were management accountants, against 134 financial accountants. There was masses of financial reporting, but no effective management information to identify the important cost elements. The standard costing system, focusing on the costs of each production/assembly process (largely people minding automatic machines, virtually idle unless there was a problem), produced meaningless information, since the significant cost lay in the (human) handling of material between processes. The costing system had to be redesigned to home in on these non-value adding activities.

How cost accounting gets it wrong

Out of date accounting has particularly serious consequences in two areas: cost accounting and investment analysis. Both of them can severely hamper a firm's attempts to come to grips with information age manufacturing. Take cost accounting as the basis of day-to-day manufacturing management. Already noted is the long-standing western obsession with labour costs even when they are three times outweighed by indirect costs, and five or six times by material;

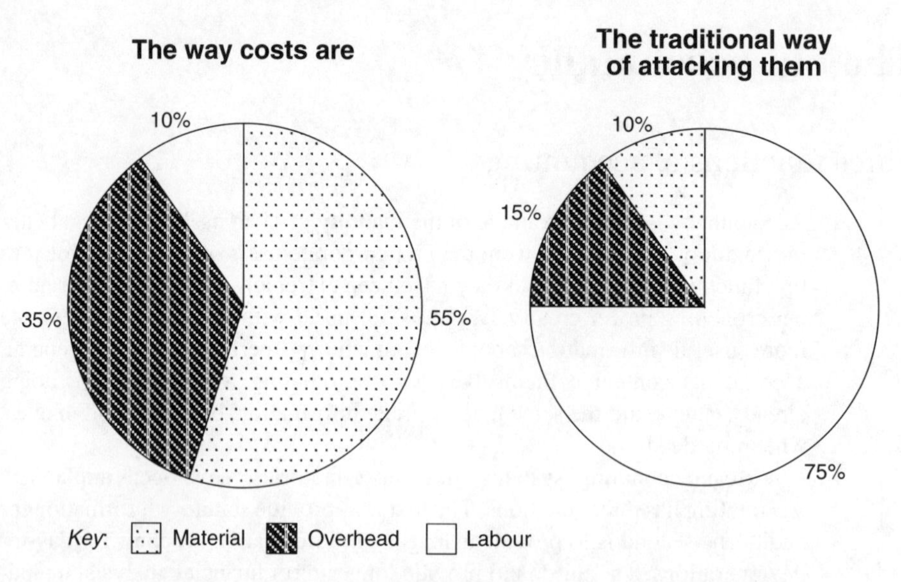

Figure 10.1 The cost problem has moved beyond direct labour costs. *Source:* Ingersoll Engineers.

even when, at the extreme, a nil direct labour cost would have no effect on the competitive position. See Figure 10.1. (At Apple's Macintosh plant in Fremont, California, the direct labour content is just 1%.)

Unfortunately, using traditional costings based on direct labour and allocated overhead to run a manufacturing business has three important drawbacks.

Firstly, to relate every other category to such a small and diminishing segment of cost 'is to erect a crazy and precarious house of cards. In a highly automated factory, the distinction between direct and indirect labour is becoming increasingly blurred. In a laser manufacturer in a West Country science park, sales engineers can be found on the production line, production engineers selling and systems engineers dividing their time between personal computers and the production line where they are installing systems in the lasers. At the other end of the scale, how variable is direct labour in a highly automated factory (always assuming it could be defined) (Sheridan, 1986). The emphasis on direct costs masks the steady movement of cost in advanced manufacturing firms from variable to fixed as the degree of automation increases, with obvious implications for the importance of extra sales and market share, and for business strategy generally. How can a traditional, variable cost, fixed output manufacturer compete with a fixed cost, flexible output operation which has levered breakeven, as in some Japanese firms, down to 30 or 40% of capacity? The answer is it cannot, however much it chops direct labour.

It is now becoming apparent just how wide and damaging have been the ramifications of the western fixation with direct labour costs. Some of the human implications were discussed in Chapter 5. There is also the vogue among US manufacturers for exporting manufacturing plants all over the globe to gain the seeming benefits of cheap labour. Not only has the advantage turned out to be transient and easily wiped out by exchange rate swings; not only has it donated lavish technological gifts to countries which would soon emerge as manufacturing competitors; it has also prevented managers from addressing the real issues involved in 'least cost manufacture', namely: quality, flexibility and speed of response.

Secondly, in the new manufacturing, the notion of standard costs is meaningless. Standard costings derive from the time and motion studies of Frederick Taylor and the days of the model 'T' Ford. Based on the simple historic measure of what it cost last time, they were undeniably valuable in circumstances of ever more precise division of labour and the mass production of a small number of standard products in huge runs. In flexible manufacturing, where the aim is to make as small as possible batch sizes of a family of products (related either by product design or by manufacturing process), standard runs are irrelevant and damaging in distracting attention from the real issues. The problem is what standard costs leave out. They ignore capacity utilization, yield, quality and whether the product is wanted at the time it is processed. All these have a large impact on 'actual' cost.

The value of the old style standard costs is further undermined by changing products and processes. Machine efficiencies no longer mean anything unless they are related to the performance on the cell, line or system as a whole. The entire sub-branch of factory economics relating to economic order quantities is irrelevant in continuous flow manufacturing with single-minute or even one-touch set-up times. Preventive maintenance becomes one of the enabling conditions for high quality and least cost, yet its value is completely unrecognized in standard costing systems.

Thirdly, as Drucker (1987) puts it, conventional accounting measures the cost of doing. It does not measure the equally important cost of not doing. It does not measure the huge cost of inventory and its associated functions; it does not measure the cost of quality (or non-quality); and it does not measure the cost of downtime. Nor can it cope with the revolutionary notion that under JIT or OPT, if at any moment people cannot add value to a product a customer (internal or external) wants, they should stop producing and do something else: housekeeping, routine maintenance, set-up reduction. Thus cost accounting ignores the important factors which make the new manufacturing work. Significantly, in Japanese plants traditional standard costing information is conspicuously absent, and few Japanese can see why it should be needed. Standard costs do not fit into their frame of reference about how to monitor and control a business. In this light it is hardly surprising if in conventional accounting terms the results of IT projects in western manufacturing plants seem disappointingly erratic. All too often, the figures used for monitoring and control measure unimportant parameters and ignore the important ones.

Accounting for investment

The potential of new techniques to transform manufacturing economics is by now well attested. As long ago as 1984 the US National Aeronautics and Space Administration (NASA) commissioned research into the benefits of CIM already achieved in five large corporations – Deere, General Motors, Ingersoll Milling Machine, McDonnell Aircraft and Westinghouse Defense and Electronics Center. Even at that stage, results included:

- 15–30% reduction in engineering design cost;
- 30–60% reduction in overall lead time;
- 2–5 times improvement in quality (measured by yield);
- 40–70% increase in productivity (complete assemblies);
- 2–3 times increase in operating time of capital equipment;
- 30–60% reduction in work in progress;
- 5–20% reduction in personnel costs (Gunn, 1987).

A year later, NEDO's Advanced Manufacturing Systems Group reported comparable gains in the UK. The report noted 'improvements in virtually every business ratio … by using MAT and, even during the period of introduction, significant improvements in annual operating profits accrue, helping to make the investment largely self-financing'. Over its eight company models, total production costs fell by 14–27% while operating profits rose 112–310%. Cash flow was negative for the first two, occasionally three, years but strongly positive thereafter (NEDO, 1985).

Such results are well supported by examples from the consultants' case-books. But there are two important catches. First, the fact that high returns are possible does not mean they are inevitable. Simplification and integration must precede IT investment, otherwise the returns will be much less or negative. Most of the gains actually originate in the reorganization phase and not with the installation of IT and AMT. Second, the investment must be made for maximum leverage. Here again there is a large invisible obstacle in inappropriate financial measuring tools which for years have channelled investment away from the areas where they would have been most effective, both financially and strategically, and into segments of lesser competitive return.

Many accountants now accept that the crude use of measures such as conventional cost accounting, return on investment (ROI) and payback are hopelessly unsuitable for justifying sophisticated manufacturing investment. They use a short-term, rule-based approach to investment which is almost guaranteed to stop the most important projects at the first hurdle or, if they do go ahead, ensure that they are undertaken 'as an act of faith' on the basis of vague appeals to 'intangible' benefits in the future. Prayer is no substitute for justification, and the facts that benefits are termed intangible says more about the people and the company's project appraisal techniques than about the project itself.

As many companies use them, ROI and payback suffer from the same overriding debilities as traditional cost accounting. They focus on tangible costs and benefits and link poorly if at all the business strategy and measures of long-term market advantage. Concentrating on cash flows over time, ROI can focus unduly on short-term profits, and Sheridan (1986) well, if unkindly, describes how managers learn to 'play the DCF game', to dire effect on the firm in the long term:

> '1. Aim to work facilities to capacity, but do not invest in increased capacity… 2. Reduce operating costs, even if this means cutting out maintenance or operating plant beyond rated capacity. 3. Do not invest in new equipment… or product development… 4. For preference use fully depreciated equipment.'

In particular, ROI and payback appraisals do not normally take account of the long-term attributes which should be the reason for the investment in the first place: shorter lead times, better quality, greater responsiveness and less waste. It is not that ROI cannot take account of these things. Rather, failure to do so reflects the reluctance of ROI analysts systematically to estimate 'intangibles', and the reluctance of decision makers to authorize projects with such intangibles providing most of the benefits.

Moreover, the combination of short payback demanded with high interest rates on the working capital has pernicious effects on the investment which does take place. By demanding a payback of 2–3 years, companies are in effect looking for a post-depreciation return on capital of 24–40%. These are levels which their normal business does not reach, and nor on average do their capital investments. It is true that most IT-based development projects can (and should) be planned to be self-financing on a yearly basis. But that is not the point of doing them. The point is at least partly strategic, and they should be justified in that light. In other words, the payback mechanism should not be used to set the hurdle rate so high that no ambitious long-term investment can jump over it.

As with most high return investments, IT-based projects typically have high risks. But the key to risk management in investment strategy does not lie in the act of faith which is now rather desperately put forward in the non-financial literature. Reports of the death of ROI, DCF and payback as appraisal techniques are premature, and to throw then out entirely because of past shortcomings in their use is undoubtedly an overreaction. In fact, precisely because the difficult to quantify long-term effects on the firm's competitive position of better quality, better design, better production control are central to the whole issue of manufacturing strategy, such benefits must be quantified as far as possible, and the need to do so is a powerful force for driving appraisal systems in the right direction. As far as risk goes, risk/sensitivity analysis is well nigh essential. Both financial and operational risks can be minimized by manual back-up and fallback procedures. Fortunately, both these issues – evaluating the pay-off and containing the risk – are largely addressed by the 'simplify first' doctrine, which starts off manually or nearly so, and proceeds by step by step implementation of increasingly risky – but highly profitable – technological investments.

The spreadsheet danger

Research has shown that over 90% of spreadsheets larger than 150 rows contain at least one significant error.

Increasingly, spreadsheets are being used to support major financial and strategic decisions. The risks and penalties of making a mistake are large and a number of companies have already lost substantial sums of money as a result of spreadsheet errors.

Large amounts of time and money are often expended in checking the input data. But the methods used in calculating the results are rarely subject to an independent review.

Moreover, the calculations performed by a spreadsheet are seldom obvious, which makes it difficult to know whether the model is working as intended.

Companies are now turning to stringent training courses which offer spreadsheet review and testing methodologies in order to improve the quality of their spreadsheet modelling.

Source: Coopers & Lybrand, 1991

New methods

Just-in-time accounting

Accounting as much as any other function must adapt to the needs of information-based manufacturing. As in the other white collar areas, there is much to be said for a just-in-time approach to accounting systems. Information gathering costs money, and more does not necessarily equal better, often the reverse. The department's aim must be to produce accounting information only as and in the quantity required: no excess reporting, no excess printouts, no waiting for the figures. Every measurement taken for accounting purposes must be questioned for its right to exist. What value does knowledge of the information add? Could that value be obtained in a simpler, less costly, more direct way, for instance by using an expert's estimate? The 'expert' might well be a shop floor supervisor, the most knowledgeable person about the area to be measured.

The response to questions like these is often that a measurement (such as inventory recording) is required for the auditors. That usually settles the matter. It does not have to. Appalled at the cost of recording audit information, the managing director of one European vehicle manufacturing plant instructed his financial director to tell the auditing firm that in the future he would record material coming into the plant and finished goods and scrap going out and nothing else apart from one annual stocktake. The auditors reluctantly accepted that this would meet the statutory requirement. The MD subsequently relented. But the experience proves that the bogeyman of the auditor, so often used to justify huge financial accounting sections and voluminous reporting, is a chimera which dissolves when exposed to tough questions about why the information is needed at all.

Appraising the appraisal system

In the same way, every manufacturing firm should submit its costing and investment appraisal systems to rigorous scrutiny. For costing, it is salutary to ask when the system was last reviewed. Does it still reflect the manufacturing realities? Unless it has been updated in the last 5–10 years, probably not. If, as in most cases, the costings are still related to direct labour, work out how much the proportion of overall costs attributed to it has changed since the system was set up, and from there how far the old categories of variable/fixed, direct/indirect remain useful. Some companies have dispensed with the direct category altogether. One snag: there is as yet no firm consensus among accountants about what should replace direct labour as the cornerstone of the new costing system. Material is one possibility that has been tried with success, reflecting the fact that it is often the biggest cost element in manufactured products. Throughput and contribution costing are other methods that have been attempted.

It is crucial to use performance criteria which link directly to the overall strategy. That is, there must be an external orientation to the figures (how is the factory contributing to customer satisfaction in the marketplace?) rather than the purely internal measurements (machine utilization and other 'efficiencies') of the present. The yardsticks used will vary according to product and processes. A line competing mostly on price needs different indicators from one that competes on high specification or quality. (In the long term, of course, advanced manufacturing is likely to abolish the implied trade-off here.) There will certainly need to be measurements of quality (defects, scrap, cost of rework, lost contribution, customer complaints), inventory and invisible inventory costs, stock turnover, productivity (measured directly, not in money terms), delivery reliability and market share. There will also be grey areas where the most appropriate costing basis is unclear. For instance, repair and service charges, and software costs, which probably did not exist when the costing system was set up. The key point is that a radical review of the costing system is almost certainly needed. As Robert Kaplan (1987) warns: 'Companies that achieve satisfactory financial performance but show stagnating or deteriorating performance on non-financial indicators are unlikely to become – or long remain – world class competitors.'

Integrated accounting

World class accounting will be appropriately integrated with all the other functions of the manufacturing firm. This is not necessarily a question of IT networks liking different departments. Rather, it is a question of both sharing information and using the resources of non-accounting areas to collect it. Thus the 'management accounts' will contain many non-financial statistics (although they all have financial implications). At another level, some of the measurements will be gathered and used by people on the shop floor. Visitors to Japanese plants are struck by the amount of quality and production information posted up in charts and tables all over the factory floor. Much of this is not strictly accounting data. But it feeds forward into the accounts, and back into the manufacturing process, thus completing the link between the two. Shop floor operators become honorary accountants, just as accountants are required to bring their function on to the shop floor.

The charts and tables underline two further important points. The first is that appropriate information technology, as in this case, may be blackboards and chalk rather than computer screens, even for accounting purposes. There is no point in computerizing for the sake of it. The second is the importance of trends. Traditional manufacturing, under pressure from corporate headquarters for short-term results, tends to emphasize fixed goals (90% machine uptime, meeting the schedule, economic order quantities) and unchanging processes. This has the paradoxical psychological effect of putting a limit on the gains expected to be achieved.

The new manufacturing approach, following the Japanese example, realizes that fixed goals can become a constraint rather than an incentive and changes the objective into a continuous daily attempt to narrow the gap between present performance and the only fixed goal that now makes any sense, zero: zero inventory, zero defects, zero delay for the customer, zero set-up times, zero batch size excess, zero bureaucracy and zero disputes with labour. The blackboards and charts, updated every day, play a vital part in this 'competitive accounting' procedure, which is constructed to support the continuous improvement of the bottom line.

Dealing with intangibles

As with cost accounting, so with investment procedures. The key is to link investment appraisal criteria to corporate strategy by making the intangible benefits tangible. This is as important in defensive cases, where the company apparently has no choice but to invest in advanced manufacturing because its rivals are doing so, as in offensive ones. Intangibles are too important to be left to chance. Examples occur in every manufacturing area.

- *Production:* lower costs through reductions in set-up times, planned maintenance (less downtime), faster response, space savings (less work in progress), fewer stock outs, standardized tooling.
- *Quality:* lower costs and improved service through integrating product and process design, and through short-cycle manufacturing; higher yields and lower scrap; improved customer service and reduced warranty costs.
- *Finance:* reduced working capital through reductions in stock and WIP and through JIT purchasing and delivery (electronic links with customers and suppliers); reduced age of debt through partnership with customers.
- *Design and marketing:* high sales through better quality, shorter and more reliable delivery, reduced quotation and design lead times and enhanced ability to design and manufacture new products.
- *General:* better use of management time as a result of simplification, integration and more effective information systems.

Beyond the figures: accounting as strategy

The new manufacturing approach demands that accountants look beyond the figures, just as product designers must look beyond products and manufacturing managers beyond the schedule. By devising the proper indicators, good accountants can help a factory to make itself world class; bad accountants using the wrong figures will almost certainly prevent it. Accountants thus bear strategic responsibility as well as tactical. More than anyone, therefore, they must be aware of the significance of the simplification–integration–automation cycle and use it to ask searching questions about priorities. They can then assist each department to use financial analysis tools such as electronic spreadsheets to quantify the options, using 'what if?' exercises. The strategically oriented accountant should be asking such questions as:

Refurbishing pulls in the pounds

The income from refurbishment is equivalent to the price of a new part, less the cost of refurbishing, less any credit which might normally be available from the original supplier for scrap. The credit aspect should not be overlooked since it might make refurbishing untenable.

In the example below, a normal credit has been allowed and added to an assumed revenue derived from other sources, the profit and loss account prior to establishing a refurbishing operation being:

		£000s
Maintenance revenue		9500
Credit for parts 'x'		100
Total income		9600
Maintenance costs:	Labour	4800
	Parts 'x'	400
	Other parts	900
Fixed overhead		2400
Total costs		8500
Profit (9600 – 8500) =		1100

With the refurbishing operation, the account will appear as follows. Note the division of the fixed overhead at 50% of labour costs, assuming no additional staff are needed to undertake the new activity.

		£000s
Maintenance revenue		9500
Refurbishing costs:	Labour	300
	Parts	50
	Fixed overhead	150
Total refurbishing costs (Parts 'x')		500
Maintenance costs:	Labour	4500
	Parts 'x'	500
	Other parts	900
Fixed overhead		2250
Total costs		8150
Profit (9500 – 8150) =		1350

Now it can be seen that the refurbishing activity by itself produces a loss, since parts 'x' are refurbished at £500,000 whereas they could be purchased for £300,000 allowing for the credit. But, in spite of this, the overall profit of the maintenance activity increases from £1.1 to £1.35 million. This increase would be sufficient to absorb some additional staff. The numbers and resulting profit depend on how the labour-related fixed overhead increases.

Source: File W. (1991)

- *Are products designed for manufacture?* At Black & Decker in Spennymoor, managers found the key to greater flexibility, higher quality and lower cost in design for manufacture. By modularizing power tool design, they found they could take large strides to automating the assembly process, with huge cost and competitive gains.
- *Have the processes been rethought to provide the most economical layout for the product?* Lucas Diesel Systems at Sudbury deliberately held off all investment until it had reorganized a traditional machine-shop plant into straight through factories within the factory. The new organization itself identified the areas where future investment would be needed, as well as some where past investment has been a mistake.
- *Does the investment add value?* This is a crucial question for IT-based projects. Many companies are now querying the value of non-value adding investment, such as: computerized shop floor information systems (simple *kanban*-type controls may be better); automatic storage and retrieval systems (AS/RS), if pull-type production methods can do away with the need for storage altogether; and mechanical handling systems generally. The Japanese are now busily ripping out conveyors in certain kinds of manned plants: far better, they argue, to position work centres close to each other so that work can easily be passed from one station to the next, inventory cannot build up between the two and space can be kept to a minimum. In a successful short cycle plant, storage typically diminishes as a proportion of factory space in favour of process areas for value-adding activities; since individual lines also shrink dramatically, there is very often space for new ones or for consolidation from other plants.
- *If it does not add value, how important is it for strategic ends?* Ford of Europe justified its videoconferencing investment primarily on its contribution to the strategy of globalization. But it took a cost and regulatory breakthrough for the cost benefit analysis to come out positive.
- *Has the Pareto principle been followed?* The guiding rule is almost always to go for, say, 75% of the total benefits in 25% of the time at 25% of the cost. Extremely hard questions must be asked about the cost of the final 25% of benefits, which are typically disproportionately expensive and have more to do with administrative/technological rounding off than with hard gains on the factory floor.

The aim is not to obstruct investment in automation or other high-technology methods. It is to ensure that all the benefits are reaped from doing it in the right way, both during implementation and subsequently in operation. This ensures that risks are kept to a minimum.

Accounting is about helping manufacturing to eliminate waste, in investment as well as production costs. Miraculously, proper cost accounting has the important side benefit of helping to establish the financial platform for capital intensive IT-based investment by 'liberating' hidden assets tied up in overheads and inventory. Lucas Diesel Systems took £30 million out of its operating costs by reorganization, laying the foundation for later high tech investment. JCB funded its investment in new products and equipment in the early 1980s by wringing money out of inventory. These are not isolated examples. Many western manufacturing firms complain that they cannot afford the high ticket items of IT-based manufacture. In most cases they do not need them, at least to begin with. When they do need them, the means of finding the money is generally under their own noses, if they can get their accountants to develop the vision to see it.

References

Coopers & Lybrand (1991). *Getting Spreadsheets Right.* An IT Partnership bulletin

Drucker P. (1987). Why automation pays off. *The Frontiers of Management.* Heinemann

File W. (1991). *Cost Effective Maintenance.* Butterworth Heinemann

File W. (1991a). The trick of turning cost into profit. *Boardroom Report: Maintenance.* Department of Trade and Industry

Gunn T. G. (1987). *Manufacturing for Competitive Advantage.* Ballinger Publishing Co.

Kaplan R. S. (1987). Yesterday's accounting undermines production. *Harvard Business Review*, July–August

National Economic Development Office (1985). *Advanced Manufacturing Technology: the Impact of New Technology on Engineering Batch Production.* NEDO, Advanced Manufacturing Systems Group

Sheridan T. (1986). How to account for manufacturing. *Management Today*, August

Skinner W. (1985). The factory of the future – always in the future? *Manufacturing: the Formidable Weapon.* John Wiley & Sons

11

The human factor

Information and people

Information intensive manufacturing has revolutionary consequences for traditional roles of people in production.

As people in the factory and offices get fewer, they become more important, not less: negatively, because there is no more just-in-case surplus to cover for deficiencies, positively, because their responsibilities increase. The Japanese view is that since for at least the next 20 years plants will depend on the interaction of people and machines, the plants must be designed round the multiskills of the remaining people, not vice versa.

Companies have always paid lip-service to their people ('our most important asset'), just as they do to customers. But the importance of people in responsive automated manufacturing is very different from in the past. Then 'participation' was an optional extra, something 'progressive' companies did to compensate workers for boring and uninfluential jobs. The new manufacturing does not work without something more than participation: full employee involvement (EI, as it is becoming known in the USA).

Involvement is essential because only people can initiate change. In the plant, multiskilled quality/maintenance engineers and setter/operators are the source of continuous reductions in set-up times and other process improvements. Senior managers do not have the detailed on-the-ground knowledge to implement or even envisage such change top down. It must come from the bottom up. The same applies in the just-in-time office. At corporate headquarters, managers must be able to use organizational change as a competitive weapon. In the 1990s this kind of management will be one of the most important arenas of competition. If technology is transferable, western companies need to be able to make better strategic use of it. As ICL's managing director has put it: 'We're aiming to make it tough [for the low cost competitors] to come in technologically; but we also have to make it hard for them to match our management systems' (Caulkin, 1987).

In no waste, world class manufacturing, the factory is only as good as its weakest link. There is no cushion of contingency. So the human organization must be world class too. All through the company, high value adding management means making organizational fluidity a normal condition of existence. At all levels, the deliberate use of change will strain traditional attitudes to human resources to breaking point.

The requirement to initiate change turns all workers into information workers; every job involves entering, interpreting and acting on factory data. This new role, with its heavy emphasis on flexibility, puts a premium on intensive learning and training at all levels. In the case of IT-based packages (automation, FMS, robotics), training will not be a separate item but an integral part of the process of implementation. In addition, training must be job related. Implementation and training must be scheduled together so that each reinforces the other. Few companies yet realize the urgency of this high level training need or are equipped to carry it out. Implementation of advanced manufacturing technology *is* training, and vice versa. Like manufacturing improvement, in the information age training and learning have no end.

A new approach to labour

Nissan vs GM

Consider two examples of high technology implementation, First, GM's car plant at Lordstown, Ohio. Opened with fanfares in 1972, this brand new automated plant was an expensive disaster. It suffered from the start from absenteeism, high labour turnover, disappointing productivity and all the other signs of chronic worker alienation. Now look at Nissan's plant in the UK. The first Japanese auto plant in Europe, it is also highly automated and quite different in structure and organization from any other car assembly site in the UK. Nissan Washington has met all its quality and productivity targets and with a $1000 cost advantage per car is reportedly the lowest cost car plant in Europe. In 1988 it decided to go ahead with a second phase of expansion and recruitment.

Lordstown was a classic case of seduction by technology. In designing the plant for machines instead of humans, GM, like many other manufacturers, missed the opportunity to turn the workforce from part of the manufacturing problem (the implicit western assumption) into part of the solution. Nissan is the reverse, an object lesson which demonstrates that advanced IT-based manufacture has to integrate the human factor as much as the technology itself. As long as there are manned plants, there are no exceptions to this rule.

In retrospect, it seems clear that the first wave of CIM-type automation was unsuccessful partly because it was simply the extreme form of the manufacturing obsession with reducing direct labour costs. Behind it was the implicit notion that getting rid of people was necessarily a good thing. The new manufacturing, on the other hand, offers a unique opportunity to recombine the jobs fragmented by previous well-intentioned job design and for the first time unite the benefits of both men and machines.

The fragmentation of the factory

Over the last half century, factory work has been configured by two conflicting currents. The first was Taylorism, which concentrated on breaking jobs down into ever smaller fragments. The systematic division of labour has brought many economic wonders, but at the extreme it founders on fundamental human needs. The production worker in the scientific management model is a factor of production and nothing more. This is the system that Chaplin effectively satirizes in *Modern Times:* not the profit motive so much as the domination of human by machines, the mechanization of humankind.

The pure work study approach to factory jobs drew a reaction in the 1960s in the form of the influential behavioural science school of human relations. Its representatives (Maslow, McGregor, Herzberg, *et al.*) concentrated on motivation as a determinant of productivity, and their investigations led to calls for job enrichment and open management, and to the vogue for participation which flourished in the early and middle 1970s. Participation was not a great success, not only because recession and unemployment, with active encouragement from pro-market administrations in the USA and the UK, pushed it on the back burner. The real reason was that the supposed beneficiaries remained unconvinced of its value. Participation did not change their status in the firm. Nor was it clear what they were participating in. 'To participate' was a verb without an object: a fundamentally patronizing attempt to give the workforce just enough say to coopt it into the unchanged purposes of the firm.

Putting the pieces together

The present unexpected opportunity lies in the fact that the new manufacturing destroys another traditional trade-off, this time between efficiency and requirements for human shaped jobs. Satisfying the needs of people at work and furthering the logic of the job are not mutually exclusive aims. On the contrary, the way to the one is directly through the other. The work-centred and the human-centred approaches to work design are reconciled by the demands of information-based manufacturing. Information technology must be used to support and amplify this tendency.

The apparent miracle arises not from technology but from the ability of the simplify/integrate phases of manufacturing reorganization to restore purpose to activity and put people in charge of real value adding jobs. Complicated methods shroud purpose in inventory, waiting time, incomprehensible routing paths and piles of computer printouts. Strip the veils of complexity away and cause is reconnected with effect, with the waste and cost of the old methods revealed for all to see.

The effect of the revelation on the factory floor is often remarkable and a depressing indictment of the way manual work has been designed and managed in the past. Indeed, in all the new world class manufacturing plants, it is the human transformation that is most striking. Time and again, managers marvel at the explosion of shop floor creativity released by close coupled organization and clear assignment of responsibilities; 'a stunning demonstration', as one said soberly, 'of the talent that we had not only been ignoring but institutionally repressing'.

Similar stories emerge from every successful new generation plant. Schonberger (1986) reported from the USA that 'Some of the assemblers and machine operators get more excited than anyone else. And why not? Who has a clearer view of the waste, excesses, mindless complexity and – from their view – bad management than those who make the product?' At Honeywell, a participant noted that as JIT was introduced, with measurements of throughput, cycle time and work in progress, 'The benefits of the system became obvious to everyone who was involved.' Commenting on the subsequent improvements in quality, he added: 'The training of operators in JIT principles and in other jobs to improve their flexibility became a two-way process which gave them an input to management and engineers as to how their jobs could be improved. This provided more ownership of the job' (Gilmour, 1987). At other plants shop floor workers have come up with solutions that managers never dreamed of: like the quality improvement group at Black & Decker which, concerned about a problem with a bought part, put the matter right by the radical means of going straight to the vendor's shop floor, collapsing about six months' delay by bypassing both sets of managers. Now such contacts are a recognized part of supplier development.

New jobs

The consequences of making purpose visible are profound. In the first place it changes the boundaries of the job. Giving someone authority to control the cause, not just responsibility for the effect, adds a new dimension to the role.

Evolution in employee skills

Old	*New*
Machine operation	
Skill is in the set-up	Skill is in simplifying the set-up
Sometimes set-up technicians or engineers needed	Operators lead projects, technicians and engineers help
Operator watches machine run	Operation is a well timed routine; operator is busy thinking about next improvement
Assembly	
Assembly jobs were simplified so unskilled labour perform them	Assemblers acquire: • multiple job skills • data collection duties • diagnosis and problem solving talents

Source: Schonberger, R. J. (1986).

Each worker becomes part of a chain on internal customers and suppliers. The job is no longer to machine a batch of parts. It is to deliver to the next station an accurately machined part at a certain time. Quick feedback on timing and quality variations identifies problems fast. Once operators understand the logic of the process, the next step, as in many Japanese and some US companies, is to give them authority to stop or slow down the line; carry out routine maintenance; and make improvements to work processes (the main activity of so-called quality circles in Japan).

Thus, functions which previously belonged to staff, particularly quality and maintenance, but also some data recording and production control, are brought into the operators' jobs. At Nissan, process improvement is part of everyone's function. At Black & Decker, multiskilled production workers routinely use techniques which were the preserve of managers ten years ago. In the past, for all managers' ritual genuflection to 'job satisfaction', they stopped short of making operators' jobs more like their own.

For the first time, factory jobs are now recognizably modelled on a similar pattern and share similar behavioural assumptions: the operator's direct function (sometimes small, particularly in automated systems) being supplemented with functions of monitoring progress (quality, inspection, data recording) and problem solving (maintenance, reducing set-up times, smoothing out bottlenecks). By shortening the information feedback loops and placing responsibility where things actually happen – on the line – is it not an exaggeration to say that the new manufacturing puts operators in a position to *manage* the resources they have to hand.

The end of participation

In turn, the new jobs make obsolete most of the terminology of recent personnel management particularly as it concerns participation. Participation in the western sense is a superficial notion. It is an attempt to generate artificial

Case study: Nissan

In his book *The Road to Nissan* (1987), personnel director Peter Wickens recounts that when the Japanese firm was setting up its much publicized car plant in Washington, Co Durham, it decided that for maximum flexibility it must have the absolute minimum number of job titles. Until 1985/86, when they were reduced to 52, Ford UK had 516 different manual worker job titles. Nissan wanted just two categories to cover all manual tasks: manufacturing staff and technicians. There were to be no job descriptions; Nissan believed that at manufacturing staff level, flexibility meant expanding the job as much as possible.

But flexibility goes beyond manual operations. Managers too must be flexible, doing their own filing, spending long hours on the factory floor and helping out where problems occur. Within their capabilities, indirect workers help out on the production line. A white collar worker might be asked to do a material handler's job. 'Once you start on the path to flexibility', said Wickens, 'there is no logical limit, other than the fact that the cost of training everyone to do everything is disproportionate to the benefits.'

Organizationally, flexibility has many more implications. Stressing the primacy of production and the supporting nature of all other departments, Nissan came to the logical conclusion that blue collar workers could hardly have inferior working conditions to office staff. So the company is single status. It also has a single pay structure based on a radically simplified job hierarchy. At Nissan, all tasks are covered by 15 job titles, from managing director to administrative assistant, and within production there are just six levels from MD to manufacturing staff. Wickens noted:

> 'Everyone at the same level in the hierarchy is of the same status. Thus there is at present only one level of manager and all are simply called manager. Equally so with engineers. Nissan does not employ production engineers, design engineers, process engineers, industrial engineers, etc – it employs "engineers".
>
> 'Of course people work in departments, but there is no organisational impediment to moving across. Especially important is the fact that engineers, supervisors and controllers [ie support professionals] are at the same level in the organisation. This both reflects the view that all of these positions are of equal importance and again facilitates movement.'

As in many Japanese companies, but rarely in the UK, manufacturing staff have complete responsibility for the quality of their work, for housekeeping and keeping their area clean and painted, and parts of maintenance. Maintenance men are themselves multiskilled, or on the way to multiskills. But the ultimate test of flexibility is the ability not just to respond positively to external change, but actively to initiate it. Nissan has developed a concept called *kaizen* (continuous improvement) in which all staff contribute to

continues

continued

improving quality, safety and productivity as a normal part of their job. Commented Wickens:

> 'This means fully involving employees in the change process so that as far as possible the person who originates an idea sees it all the way through to completion.
>
> 'This may result in the individual concerned preparing a drawing, rewriting a process sheet and actually undertaking the physical change. The Nissan body construction shop is now considerably different from the original layout, and the vast majority of these changes have been thought up and implemented by the people working in the area. In many environments management fights to introduce change whereas in the Nissan environment the individual "owns" the change, and the problem is not introducing change but keeping up with the proposals.'

responsibility for management tasks among workers who are untrained and unsuited for such roles. Participation in Japan is so endemic to the normal working environment that many Japanese commentators find it difficult to describe in terms comprehensible to western ears. Participation to them means the involvement and the 'turn the hand to anything' flexibility which is the normal condition of work in a very small firm. It stresses simple linkages between processes, visible methods over black boxes, clear responsibilities for quality and timely delivery and a combination of staff and line duties for everyone. By doing the same, the new manufacturing can build back into factory jobs the meaning which generations of manufacturing engineers misguidedly took out. Participation no longer needs to be designed into jobs since it is already there, with all the requirements which that suggests for different attitudes on the part of both managers and the managed.

Convincing the managers

The myth of shop floor resistance

Managers often cite shop floor resistance as a reason for not introducing new generation manufacturing methods. The issues of single union plants and no-strike deals are not dead. Established plants must negotiate flexibilities across a multi-union structure. If they follow the logic of single status, as adopted by Nissan, there will be delicate issues of harmonization (although perhaps not all as intricate as the multi-union Alcan Sheet plant quoted by Wickens (1987), which found that 'over the years it had established four holiday plans, five job evaluation systems, seven wage and salary structures, three working weeks and six overtime arrangements).

It is also true that people will certainly be suspicious of old style, technology-first investment, as at Lordstown. And new organization and techniques have to be prepared, trained and paid for. But no world class manufacturer cites people as the problem; rather the reverse. Once a pilot investment has gained experience and commitment, the benefits are usually so obvious that resistance is minimal. In some cases the problem is keeping up with suggestions for change rather than initiating them. 'The only thing people resist is stupidity', summed up the managing director of one successful high tech plant. Said another: '90% of problems aren't to do with resistance, but not being explained. When communications are so much better, they just disappear.' A third: 'I never met anyone yet who would do a bad job if the tools were available to do a good one.'

Much more problematic is the attitude of managers.

> 'Five years of experience in more than 20 factories in the USA and Europe where JIT has been implemented has led to the belief that the [poorest] resource in many factories is not the work force but more usually the middle management. This is especially so of production engineers, who are fewer in number than their Japanese counterparts, and who appear to be completely discouraged between a stubborn work force and the short term profit oriented top management decision making policy.'
>
> (Abe, 1986)

A study in the UK concluded that 'The greatest barrier to the further use of IT in industry and commerce is management itself' (Kearney, 1984). In practice, a major change in attitudes, organization and behaviour is needed at all three levels: shop floor, middle and top management.

Simplified management structure

In the first place, the new manufacturing makes both horizontal and vertical segments of management redundant in its present form. In the 1960s and early 1970s staff headcount in many western companies ballooned as planners and computer people were taken on in the attempt to manage complexity. Layers of management grew with the staff subempires until some US companies found they had built high-rise management edifices up to 15 levels tall, double the number in their most efficient Japanese competitors. At the same time, support people proliferated around the factory floor in quality assurance and maintenance departments, together with accountants and data processing staff who were brought in to wrestle with bulging inventories.

Driving the increases was the assumption that data was scarce, and that more data permitted better decision making. Now that data is plentiful, largely because of the ubiquitous computer, the need is to turn it into information by giving it purpose. Here the logic is reversed: economical use of data permits better management. Multitiered managements hinder rather than speed the flow of information. They are also extremely expensive. Conclusion: some layers have to go, as they already have done in companies which felt the pressure of Far Eastern competition. In the USA alone, half a million middle managers have been pushed out over the last five years, with more to follow. One US company described in a NEDO report (1985) is now run with 'no managerial hierarchy, no middle management, no first line supervisors and no quality control units', but a lot better use of technology and improved overall productivity.

Simplified manufacturing organization means simplified management. JIT-based factories find that since in close-coupled systems the process itself decides what happens next, fewer manufacturing managers are needed, and those that remain can concentrate on strategic goals (quality, process improvement, training) rather than firefighting. At Lucas Diesel Systems in Suffolk, reorganizing the plant into focused minifactories meant that the plant management hierarchy was reduced from seven layers to three, with enormous gains in accountability and problem resolution. Black & Decker's power tool cells now need 20 manufacturing managers to run them, compared with 68 a few years ago.

Proper information techniques ensure that these sweeping simplifications strengthen rather than weaken management. In today's terminology, the old structures were supporting not the management of value but the management of waste. The aim of the new organization is simple: better maintenance with fewer people in plant maintenance, better quality with fewer quality controllers, better accounting with fewer accountants, better production control with fewer production managers, better materials management with fewer materials staff and better information with less data processing, as Schonberger (1986) sums it up.

Case study: H. J. Heinz

The major food manufacturer, Heinz took on the task of bringing its workforce and their working practices up to the quality level of its high investment in new computerized production equipment.

continues

continued

This was a huge undertaking; Heinz has a 3000 strong workforce in the UK, split between three production centres, so a pilot programme was launched at the 150-strong pasta production centre in north London.

Communication was the key factor. Management spent four months explaining the reasons for the changes to seven different union group representatives. Agreement was reached to retrain the workforce to increase flexibility and accept new ways of working.

Apart from training the workers on the new production equipment, a major reorganization took place. The employees were put into self-sufficient 'cells'. Each cell completed all the processes on one product group. The multidisciplinary team were trained in a total quality management approach and encouraged to identify and solve problems – working as a team became all important.

Electricians and fitters who had previously worked as a separate unit, responsible to their own foreman, were split up and allocated to the different production cells. Each cell became responsible for its own production throughput, safety and maintenance. The increase in levels of responsibility caused some problems for some people. Selection of individuals to make up the teams had to be done carefully, taking into account those who were happy to take on extra responsibility and those who were somewhat afraid. The 150 staff spent at least four weeks in team training, learning how to function as a team and building up their confidence to contribute to the decision-making process.

After two years, the pasta centre has fulfilled its long-term corporate objective and has improved its running capacity considerably. Such a radical change in working practices, however, requires ongoing monitoring and reinforced training. This has continued successfully throughout the Heinz organization.

Working together

Managers are having to learn to work more flexibly. Just as the new manufacturing breaks down the demarcation of traditional craft skills, it also demolishes cosy departmental divisions. In the past, designers lobbed a new product over the wall to see what manufacturing and sales could do with it. Incorporating changes and repeating the process produced long design to production lead times and complicated products. Design for manufacture (actually a crucial part of pre-automation) demands that all these inputs are coordinated very early on. Product designers must work closely with production engineering and spend time both on the shop floor and with the sales and field service teams. Sales people need to know about manufacturability and its relation to just-in-time delivery.

Job rotation, at present a rarity in western companies, will become much more prevalent, at the expense of purely professional loyalties. IBM rotates all managers as part of its meticulously developed employee policies, emphasizing learning as an investment rather than a cost. So do almost all the large Japanese firms. Wickens (1987) noted:

> 'The Japanese administrator, who joins from university, will frequently not know into which department he will be placed and will during his career expect to move from one function to another. The engineer is less mobile, but with the Japanese emphasis on the closeness of engineering and production (which are often indistinguishable) he too will have a broader career that his Western counterpart.'

The 'pure' specialism is not highly regarded in Japan.

Common to the new disciplines is the emphasis on the shop floor. Engineering knowledge is the baseline requirement. By stripping out the non-value adding activities which have traditionally provided comfortable staff jobs for managers, companies can concentrate all their attention on the two things that really matter: manufacturing effectiveness on the shop floor and getting close to the customer. The days of the specialist staffer, working in tranquil isolation from the shop floor, are numbered. Flexible manufacturing demands flexible managers, strong on teamwork, unpreoccupied with hierarchical status and creative in devising new organizational solutions for problems as they arise.

Preparing the board

High technology of all kinds poses special problems for senior management. As with any new technology, initially only specialists know about it. For the untechnical majority, there is the conumdrum of how to find out about what they are not aware of (and when they are aware of it, how to learn without losing their pride). The result is the common and disastrous tendency of IT-based projects to be technology led rather than business led. A second set of difficulties concerns implicit assumptions about change. Although most managers profess to accept the notion of permanent change, in practice they operate on the belief that the more continuity and predictability the better. 'Better than last year' comes to be the yardstick against which new proposals are instinctively judged, and 'if it ain't broke, don't fix it' the unspoken motto.

But these are not good game plans in present conditions, where internal measurements are insufficient and change must be welcomed. Short-cycle manufacturing embodies fundamentally different philosophies from those most western managers grew up with. IT adds a further dimension which takes operations into little known territory. It is often hard for senior managers, who have succeeded in very different circumstances, to take on board the new manufacturing realities, particularly when they are not themselves technically minded or used to learning completely new concepts affecting the way they work. In too many cases they abrogate what should be business responsibilities to technical specialists.

There are three partial solutions to the boardroom problem. One, appoint a top level technology custodian to oversee not just product R & D but process improvement as well. Studies of change all tend to show that pushing over barriers of vested interest and scepticism requires an enthusiastic change agent to keep the momentum going through difficult periods. (This is as true of organizational as of technological change.) Two, commission a technology audit to scrutinize the company's products, processes and R & D. The investigation is best undertaken by an independent third party who cannot be accused of departmental partiality or of favouring top management's preferred solution. For a company setting out to overhaul its manufacturing strategy, a technology audit can prove an invaluable source of baseline information. Three, send them back to school.

Training for new technology

Three reasons to train

There is no disgrace in educational upgrading for senior managers. On the contrary, there is no doubt that planning for management training and retraining at all levels will be one of the key elements separating the successful from the unsuccessful companies in the age of IT. There are three reasons for this. The first is technical. A strategic view of technology is essential, and the more complex the technology becomes, the more important the clear vision. Flexible, real-time factories are different from the processes managed in the past. The creative leadership demands of the flow-state organization are quite distinct from the requirements of day-to-day inventory driven manufacture.

For management, the engineering input will be paramount. If Japan, which graduates proportionately six times as many engineers as the USA and the UK, is currently experiencing a shortage of this kind of management expertise, how much more urgent is the need to boost both quantity and quality in the West? Wickens (1987) again:

> 'Because of their education level, the seniority system, continuous in-house training, development and rotation and sheer dedication, the general level of detailed technical knowledge of senior Japanese management far exceeds all but the very highest in the UK, the USA or West Germany. Nissan Manufacturing UK recruited some of the best production and engineering management from the British motor industry and they have all been amazed at the strength of their Japanese counterparts.'

Without a clear understanding of these issues at corporate level, there can be no systematic review of the training needs of the rest of the company.

The second reason for training is organizational. Managers, particularly older managers, are not used to thinking creatively about organizational change. Yet this will be their chief contribution to the competitive company of the 1990s. While shop floor engineer/operators are working on continuous improvements in process, managers will be constantly seeking to use organizational change to keep ahead of the opposition. A few companies already use constant change to weed out less adaptable managers and encourage them to move elsewhere. As technology becomes more and more transferable, the premium shifts to competitive elements that are harder to copy, notably the human organization that can consistently deliver quality and reliable service. ICL's commitment to competing through its management systems has already been noted. At a time when the company's product line is under attack, IBM's accumulated investment in training for quality and service is one of its most powerful competitive weapons. Other companies will need to consider how to deploy their organization as a weapon, and train their managers to use it.

The third reason for training is reassurance. Installing any IT systems can be a vastly difficult exercise at the human level, cutting against the grain of long-established group loyalties and cultures and substituting a wilderness of open-ended roles for the familiar organizational territory of the past. The psychological effects on individuals can be dramatic and disturbing. Significantly, a study of the manpower aspects of new manufacturing technology found that 'North American companies reported repeatedly how great the resistance to change is, even in their environment' (NEDO, 1985).

It is important here to distinguish between the methods of advanced manufacturing, such as just-in-time and quality techniques, and the technological tools. It is the tools which are potentially alienating. It has become clear with experience that JIT and TQC, particularly used together, have strong motivational qualities in themselves. Good manufacturing practice, as it is now understood, is integrally motivating. The happy consequence is that, by training for JIT, companies carry out much of the human groundwork for the later introduction of advanced technology. Once again, the systematic approach to low technology is the essential precondition for high tech solutions.

Multiskilling means flexibility

One of the greatest problems in training or retraining employees, highlighted in several surveys conducted by consultants, professional bodies and unions, is the way in which those skills are put to use after the training period. Frequently, it would seem, companies do not make proper use of the new skills acquired because they have not restructured their organizations sufficiently to have removed some of the old lines of demarcation, or they have not developed a proper business strategy which highlights the specific training needs in relation to the development of the technology and the organization it supports.

The maintenance example

A short-sighted view of training is perhaps most easily identified in the area of dealing with breakdowns and preventive maintenance. According to Lynn Williams of the UK's Electrical, Electronic, Telecommunications and Plumbing Union (EETPU), 80% of breakdowns are relatively simple and could probably be rectified by the first available multiskilled craftsperson. She argues that a versatile, multiskilled craftforce can effect simple diagnostics and fault determination to reduce delays, leaving the core specialists to deal with more complex breakdowns.

In the EETPU's experience, when a company embarks on a multiskilling programme, it often makes the mistakes of not defining a start point, being overambitious in its expectations and, by not communicating effectively with employees when developing a training strategy, not eliciting valuable input which could dramatically affect the eventual outcome of any drive to achieve multiskilling. The union suggests that the evolution of maintenance skills involves a constant advance towards future technology. It starts up the training ladder at the core skill level. The first stage involves developing versatility between skills. When this is achieved, progress is made towards maintenance teams and systems training. This creates a more efficient and effective maintenance workforce which can deal with sophisticated plant and equipment. A meshing with production skills is inevitable. Indeed maintenance production cells that feature teams of systems technicians hold the key to effective manufacturing in the future (Williams, 1991).

The competence of operational/production managers

Several of the reports quoted in this book appear to have found evidence that the development of operational and production management skills has been seriously neglected in some areas of Europe. One recruitment consultant that specializes in pan-European executive placement said in 1992,

> 'As companies become better at analysing their exact corporate needs and planning their strategies, so they become better at knowing

exactly what skills they require of key personnel. We recently took an unprecedented five months to find a production manager from a Belgian brewing company because we could find no-one with the right skills who was actively seeking employment. In the end we had to headhunt, a practice we try to avoid and one which we are conscious does not add to the available skill base in Europe. From our point of view we have seen a deterioration in skills available at this particular level of management.'

The DTI has published a very lengthy checklist of what it considers are the skill requirements for operational production managers, under the broad headings of People Management Skills; Broad Operations Management Skills; Information Analysis Skills, Business and Financial Skills and Change Management Skills (DTI, 1989). Clearly this is quite a parcel of knowledge for one person to handle but the evidence is that world class manufacturing cannot function properly without a high level of skills among *all* of its personnel.

The dangers of offshore programming

A recent development in the IT industry in Europe is causing waves of alarm to spread through training and recruitment specialists. Companies who seek to cut the cost of their monthly IT wages bill are increasingly using the services of offshore programming specialists in an effort to slash the cost of software development. Experts consider it to be a short-term, short-sighted response to the economic climate that may, eventually do a great deal of damage to the skill based within European countries. By stepping over a generation of IT professionals and sending the work to the Third World and Eastern Europe, the IT related professional bodies within Europe claim that it will denude the job market for the next generation of IT professionals that are being encouraged to train right now.

Learning curve options

The variety of technology on the market has, in the past, led to a variety of training methods being offered. Proprietary courses by the various manufacturers have always been on offer but the increasing drive towards open systems has led to the development of open and Unix training facilities. The

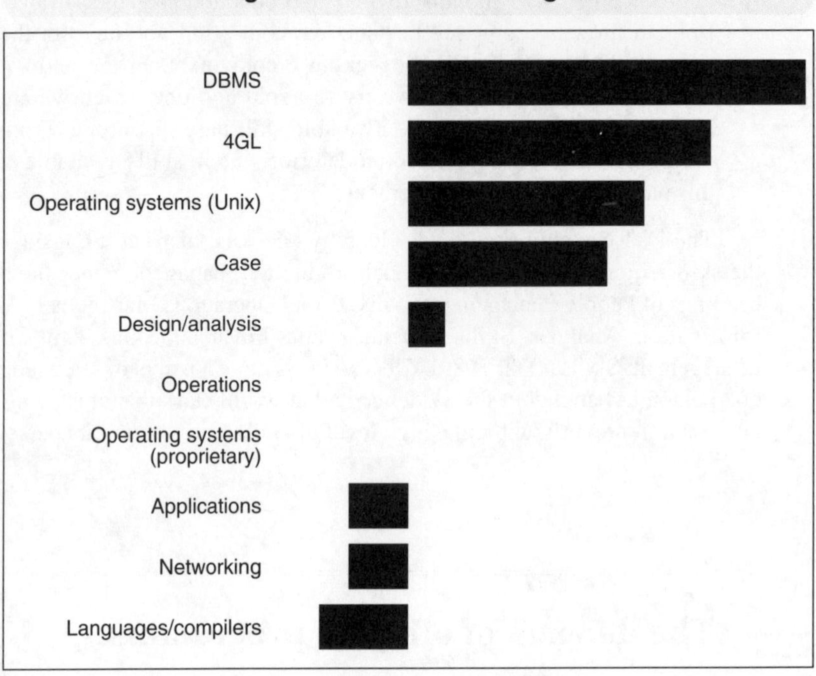

Figure 11.1 Annual growth – IT skills training 1992–95. *Source:* IDC, 1992

demand has grown at approximately 8% a year in the UK since 1992, according to a report by IDC, although the same report found that demand for training in database management systems and fourth generation languages were still in greater demand (Figure 11.1).

The recession and the move in general towards Unix-based systems meant that companies that plan to come out of the recession with the right skills have continued to invest in training.

Computer manufacturers have seen the light too. Unisys invested over £12 million in a purpose-built residential training centre in England and offers Unix as well as Unisys proprietary courses. Several of the manufacturers have residential sites although predictions are that technology changes so fast that companies no longer have the time to send staff to residential courses.

Hence the development within the training market of computer based training (CBT) or adjusted time learning as one manufacturer, Digital, calls it. This is a flexitime method of learning, either at open learning centres, most popularly as evening courses, or in-house, which could be the way of the future.

Digital uses a number of open learning formats. These include interactive video (IV), text-based instruction (TBI) and audio visual (AV).

IV uses a combination of sound, video, graphics and text to create an interesting and encouraging form of learning. Access is controlled through the interactive capabilities of the computer. The user interacts with information held on videodisk by responding to output through the keyboard. The response is then used by the software in order to decide what to reply.

TBI relies upon a learner guide (the text) as a primary route of direction throughout the course. This is often supplemented by access to job-related reference manuals. Practical exercises are also built into the text to reinforce the learning and enable individual progress to be measured. Some text-based courses have exercises and examples which are pre-prepared as data files on magnetic tape. The tapes are ready for loading onto the computer systems to assist and re-enhance the learning process.

AV is a video based method of learning that uses dynamic and colourful graphics to give an indepth presentation of a variety of concepts. Many AV courses have work books which provide further information, summaries and tests.

Another, welcome, development in the training market is the addition of 'business analysis' before designing a training course. IBM appears to have pioneered this approach, stating that they work with customers to identify business objectives and analyse the training requirements that stem from those so that training can be focused more effectively. IBM believes that people are looking for ways to assess how far training has helped them meet their business objectives.

Training in tandem with strategy

P & P Training Services reinforce this view. In an article written in 1990 the company stated:

> 'Managers need to understand that every training programme should be thought of as having two distinct functions. Certainly, it must achieve its primary objectives of increasing the productivity of each individual. But, in addition, training has an important role in reinforcing the organisation's commitment to its own IT strategy. To build tuition courses that focus solely on the practical and explanatory level is to miss the wood for the trees. As training programmes are devised for personnel at every level within the organisation, so there should

be a component within each course that is specifically designed to communicate an overall strategy and to demonstrate how the new technology and the training courses themselves provide a "win-win" situation for both the participant and for the organisation itself.'

Case study: NEDO

The NEDO study group (1985) reported in the following terms on an unnamed US company which had recently opened a highly automated green-field plant.

> 'It is the company's view that every individual has to recognise that with the new technology there are no longer clear job boundaries, and the jobs themselves are of a much more interactive nature. Consequently there is a need not just to reappraise specific job skill requirements, but also to enhance relational skills. The company believes that successfully resolving these twin skill, job diffusion and relational aspects does assist companies to introduce new technology in a significantly shorter time than would otherwise be expected.
>
> 'But the main point is that the company believes that to achieve this not only must people change their roles but that this can only be done if the new technology is matched by new organisational approaches.
>
> 'In this company there are no management hierarchies, there are no middle managers, there are no first line supervisors, there are no inspectors and there is no quality control unit. The company runs on a team A and B basis, with A providing technical guidance in consultation with B, which handles the production process. Remuneration is on the basis of skills possessed and not on work performed; there is no hierarchy within the team and no team leader for the group which numbers between twelve and 20 people. The team recruits its own personnel (after a five stage screening procedure which looks closely at a candidate's interpersonal skills) and has the power of dismissal. It is believed that this approach makes for better decisions and the most effective use of the technology and the best productivity performance.
>
> 'The company believes that while introducing programmable automation by itself can lead to worthwhile productivity gains, without new organizational approaches at least 40 per cent of the realisable benefits will be lost. The technology, it is believed,

continues

continued

demands that job divisions are removed, that jurisdictional issues between such as draughtsmen and engineers must not exist and that the prevailing philosophy must be, "Relate and participate". It also recognises that these precepts must be constantly reinforced by a process of effective communication and a continuous process of training and retraining.'

Paying for flexibility

There is another side to teamworking, multiskilling and increased flexibility: how to pay for it. Since skills at all levels are vital, experience must be properly recognized. IBM rewards people well for their new responsibilities, as do Nissan, Yamazaki and a variety of other top class manufacturing firms, with a single scale of rates and conditions for both direct and indirect staff. There are three issues here, all of which need careful thought.

(1) Basic pay for working in a multiskilled, 'responsible' environment. Flexible working allows much simpler pay structures. Many companies have torn down elaborate job evaluation systems and now pay according to skills rather than a rate for the job. Grading systems reflect the emphasis on skills.

(2) Incentives. There is no place for traditional piecework incentives in a close coupled environment, where too much production is as bad as too little. Incentive payments must therefore centre on wholly different attributes from the past: extra skills, process and quality improvement, and plant profitability are all used with success.

(3) Career paths. Breaking down departmental barriers so that managers can work in teams obliterates old career paths through the functions to a place in general management. New ones must be marked out which reward the new methods of working, not fight against them. Many of the career implications of new manufacturing methods are only just emerging. Flexibility in rewards will be needed to accommodate them.

Integrated human relations

In the no waste economy, every resource must be used to the full. Human bottlenecks in production can no more be tolerated than machine ones. With no just-in-case surplus to hide behind, each manager and operator must operate as flexibly as the new machines and with 100% attention to quality. But these imperatives provide an opportunity, not a threat. Until recently, despite the efforts of the human relations school, the progressive fragmentation of factory jobs and the resentful reaction of disenfranchized operators formed a dismal spiral of second class treatment and second class performance. The new manufacturing, on the other hand, gives ordinary performers at every level the chance to shine. As Schonberger (1985) pointed out: 'Operator centred maintenance, operator centred quality and operator centred data collection and diagnosis are not just talk: WCM [world class manufacturing] *demands* their use.' Toyota's JIT guru, Taiichi Ohno, said the same thing more tersely: 'Improvements of worker's operation first, then everything follows.'

The principles of the new human relations requirements are not hard to grasp. But there is no short cut to the flexible qualifications and structures that will make them work. Look at Japan. Japanese companies have taken 30 years to evolve by trial and error the elaborate personnel departments and policies which complement their production methods. It is often forgotten how much influence such personnel departments wield, and how important their part is in the overall company performance. Some of their meticulousness must rub off on their European counterparts if the latter are to gain the full benefits of copying their production techniques.

Case study: Neglect of human resources

This case, which featured first in a 1986 study by Clegg and Kemp and then in another study in 1989 (Willcocks) shows how neglect of human resources issues at an early stage of a computer project can show up as industrial relations problems surrounding inplementation and operation.

In an electronics company several millions of pounds were invested in a flexible manufacturing system. However, the introduction of automated systems caused a series of disputes between managers and unions, between departments, and between the various unions themselves. The main bone of contention was the failure to consider the employees and the change of work organization during the design of the system. This caused long drawn-out disputes between engineers, machine operators and programmers over responsibility for routine edits of program tapes.

The problems were caused by the FMS project being allowed to be controlled by technical specialists. Line managers, personnel department, supervisors and industrial relations specialists were not asked for any input. Human resource issues like responsibilities, working relationships and training were only considered after the system was up and running.

References

Abe K. (1986). How the Japanese see the future in JIT. In *JIT Manufacturing* (Voss C. A., ed). Proceedings of the First International Conference on Just-in-Time Manufacturing. IFS publications, Springer Verlag

Caulkin S. (1987). ICL's Lazarus Act. *Management Today*, January

DTI (1989). *Manufacturing into the late 1990s*. Department of Trade and Industry

Gilmour D. (1987). Just in time in action. *QA News*, June

IDC (1992). The UK IT Training Market – opportunities in a time of change

Institute of Directors (1990). *Training: The Basic Essential. A Director's Guide to Information Management*

Kearney A. T. (1984). *The Barriers and Opportunities of IT – a Management Perspective*. Institute of Administrative Management and the DTI

National Economic Development Office (1985). *New Technology: Manpower Aspects of the Management of Change*. NEDO

Schonberger R. (1986). *World Class Manufacturing*. Free Press

Wickens P. (1987). *The Road to Nissan*. Macmillan

Willcocks L. (1989). Information technology and human resources management: in search of strategy. London: City University Business School Working Paper

Williams L. (1991). Laying the ghost of demarcation to rest. *Boardroom Report: Maintenance.*

12

The way ahead

The paradox of technology

The post-technological age

In one sense, the technological shape of the 'business of the future' is unimportant. This may seem paradoxical, given the emphasis placed on the complexity of current solutions and the immediate technical problems, notably interconnection and software, which still need solving to get the best out of existing manufacturing hardware. CIM may not yet exist except in the minds of systems visionaries and optimistic hardware salespeople. Yet it is clear that in isolation all the individual enabling technologies for an entirely computer integrated operation are already in place, or will be very shortly. The key adjustment that companies have to make is this: they are no longer operating in a world of technology scarcity or rationing, but one of technology plenty, a post-technological age in which processing power is a commodity to be attached wherever it is needed. Modern IT permits an almost unlimited range of possible manufacturing architectures.

These are three variations on a theme of focusing technology on people which will be played with increasing frequency by the most sophisticated companies in the new few years. They underline that what matters is not (with rare exceptions) the technology itself, since eventually everyone will have it. (Hewlett Packard and a Chinese steel company, for example, have already constructed the first computer integrated plant in China.) The best appropriate technology will be necessary, but not sufficient, to make the manufacturing

Case study: Visions of the future

In Japan, **Fanuc** has built a series of 'village blacksmith' factories. Located in remote countryside, the factories are small, containing a few machine tools and robots and just two employees, both engineers. They are also cheap, cost ing less than $500,000 each. The aim is a whole network of such small scale, automated plants, making items like electric motors and products for electric spark machines (Gunn, 1987).

Volvo's Kalmar factory is a car plant without a transfer line. Since 1974 Kalmar employees have been building vehicles in teams in a specially built two level plant, each group responsible for its own results. Assembly is carried out on AGV carriers, controlled by a central computer. Kalmar's group assembly techniques have often been written about as a successful labour relations experiment. But its production record is also first class. Between 1977 and 1983 it reduced total manufacturing time per car by 40%, cut defects by 39%, increased uptime from 96 to 99% and boosted annual inventory turn from 9 to 21 times. Kalmar has the lowest assembly costs of any Volvo plant; it also shows the company's lowest white collar man hours per car produced. According to a report by Agurén *et al.* (1984) on the plant in 1984,

> 'The most interesting aspect of the Kalmar plant is the way that new technology and new organisational patterns have been combined to create an entirely new type of working environment, which has made it possible for every employee to have meaningful work, personal involvement in his day-to-day activity and a high degree of job satisfaction… The Kalmar plant is a dramatic demonstration of the feasibility of achieving extremely positive production results in an unusually favourable working environment.'

Volvo has built a new plant at Uddevalla where car assembly has come the closest yet to building on the spot, with operators standing still and components coming to them. The factory has been built with six separate product workshops, each equipped to build a complete car, rather than simple sub-assembly.

Linn Products, a small Scottish manufacturer of upmarket hi-fi, likewise has no production line. Linn decided that the whole ideal of the production line was antithetical to its aim of building the highest possible quality into its very wide range of products, from tone arms to record decks to amplifiers and speakers. Linn's radical solution was to build a completely flexible factory round a self-developed automated handling system. From the automated storage area, AGVs at the start of each shift deliver not only the parts needed for each production worker to build and pack a complete product, but the components of each work station as well. Flexibility is achieved not by buffer stocks – there are none – but by the close coupling of the delivery system and by response times measured in minutes rather than the usual hours. The next

continues

continued

stops on Linn's route to 'real-time manufacture' involved locating low level automated assembly work in the stores, so that complete subassemblies deliver just in time to product builders, and 'roving robotics' (flexible robots that plug into the flexible work stations).

company a winner. It is the elements of technology that a company chooses out of the range of possibilities, as at Linn, Volvo and Fanuc, and how it uses them to help people to add value that are crucial. In a curious fashion, the more wonders technology can do, the less interesting it becomes in its own right. The more it can be taken for granted, the more manufacturers can stop fretting about computers and concentrate on the real issue, which is making better products and winning and keeping customers. The medium is not, after all, the message. As the manufacturing director of one world class UK company sums up: 'Computers speed the flow of information. If that helps you make better products, fine. If not, spend the money on something else.'

The new message

This is a radical and positive message, sharply different from the technology worship implicit or explicit in so many official and consultancy reports. Its principal subtheme – integration – is neatly summed up in two conclusions reached by the study group which reported on Kalmar in 1984 (Agurén *et al.*, 1984). The first is that since production technology is largely determined by the product and its design, future manufacturing breakthroughs will depend as heavily on product as on process reappraisal. The point needs stressing. In the information age, product and process are intimately connected. Thus, apart from competitive manufacturing, a firm needs attractive products that people will buy. Attractive products are smaller, lighter, higher quality, longer lasting. These are all production technology problems as much as design engineering and marketing problems. A key element in the world class performance of Black & Decker at Spennymoor was a design for manufacture exercise which enabled it to modularize production, and as a result improve quality, achieve greater variety and automate assembly. Result: simpler, more cost-effective manufacture of better products in a complex multiproduct plant. In this light, the difference between a pre-automation Black & Decker drill and the current model is due to better use of information (about product usage, customers, markets) that has always been available but underutilized; just as the difference between the preorganization Black & Decker plant and the present version is the way it manages and integrates product/process information.

The second concerns the relationship between technology and people. A striking characteristic of all world class manufacturing firms is the way they use simple organization to liberate the hidden human resources which are the foundation of all lasting competitive strength. By adapting work organization to people and making new forms possible, new technology can play a significant role. But technology should be subservient to, and a significant enhancer of, people and products, not the other way about. In other words, this is a people-and-product-centred approach to technology. To look at technology as a substitute for people and a cheapener of products, as many companies have done in the past, is essentially negative. Its far greater potential is positive and as yet largely untapped: the opportunity it affords to realign a company's structure and processes with the demands of low cost, high quality and increasingly responsive manufacture.

Case study: Octavius Atkinson & Son

A £15 million investment, of which £7 million was used to install the latest in steel processing machine tool technology, has made Octavius Atkinson the most modern and efficient structural steel manufacturing plan in Europe.

During the 1980s, the company had a market share of nearly 25% and a turnover of £30 million. It decided that it was time to invest in the future.

The new factory, a 20,000 m^2 facility at Flaxby, reflected the company's new manufacturing philosophy. Automated production replaced labour intensive methods. Plate and sectional steel parts up to 60 tonnes can be lifted and processed automatically through punching, drilling, shearing and bending lines. Computer-aided design is used extensively, and off-line machine part programming by direct numerical control has been installed to further speed throughput.

This technology is supported by a two-year programme of investigating and developing new markets in Europe, as well as a continuing and successful marketing exercise in the Middle East and the US. The company's philosophy is that technical excellence should go hand in hand with market expansion. It is a business strategy based upon maximizing the confidence created by investment in technology.

The elimination of waste

A strategic view

The best manufacturing firms in Europe and the USA are now reassessing the route their factories have travelled in the past. Under the weight of competition, they have been forced to acknowledge that while they have been sweating to reduce labour and control complex plants through the use of increasing doses of 'technology', a new and more productive line of manufacturing advance has passed them by. It has taken the success of Japanese companies based on a rigorous philosophy of eliminating waste in cycle time, inventory and in poor quality, to reveal a different and broader perspective of manufacturing change. Hardware advances (NC, CNC, CAD/CAM, CIM and others yet undreamed of) simply mark the stages in a never-ending stream of strategic business improvement from low cost to quality and now to increasingly flexible, responsive production.

In the face of this stern doctrine, western companies badly need a carefully thought out strategy for using IT to guide them back to the right track and show them how to separate the enabling from the disabling technology. For the reassessment of technological tools is the other recurrent theme of this book. Most books and all suppliers' literature emphasize the potential advantage to be gained in manufacturing through the use of computers and IT. In so far as the analysis goes, it is true that they can create competitive edge, although since little of the technology is proprietary, the lead is often short-lived. But few people have pointed out that IT is a double-sided coin. The bright uppermost face is eagerly discussed, although it is surprising that after so many years the case histories of manufacturing companies using IT to win lasting competitive advantage are remarkably few. The dark underside of accumulating cost and overheads wins much less attention.

IT as a burden

Yet the evidence for the burden of inappropriate IT is ubiquitous. Ironically, much of it occurs in reports and surveys which enthusiastically ignore the disagreeable implications of their own findings and argue that the answer is yet more technology. More or less at random, a survey of corporate staffs in the USA found that, over all industries, information systems was the largest single headquarters function with 23% of total employees and that chief executive officers were increasingly concerned with the difficulties of managing IT (McCormick and Paget, 1987). A report in the UK suggested that more than half the at least £1 billion spent annually by British industry on software was wasted because of the poor quality of the product (DTI/Logica/Price

Waterhouse, 1988). Another British report underlined that many senior managers do not understand what their firm's computers do and spend much of their energy disguising their guilty secret behind a 'culture of bluff' (Coopers & Lybrand/BIM 1988).

In addition, *Business Week* (May, 1988) lamented 'the widening gap between hardware and software performance' and noted that with the length of programs increasing by 25% a year, computer aided software engineering (CASE) was set to become a $2 billion market by 1992, the cost of which, of course, would be paid by the eventual end user. In 1984 the IAM/DTI/Kearney study of IT barriers and opportunities found that UK companies were wasting an average of 20% of their total IT spending through such failings as misallocation of resources, disregard of customer service and overengineering or overspending (Kearney, 1984); looked at another way, 50% or more of all information systems installations are commonly regarded as failures (Lyytinen and Hirschheim, 1987).

Given the negligible effects of burgeoning computer spending on overall productivity, the infernal nature of the computer created machine becomes clear. There is, alas, no such thing as a technological free lunch. Processing may be getting cheaper by the minute, but every extra notch on the ratchet of technological complexity beings its own inexorable overhead cost: in information storage (IBM built a massive new office block at is Montpelier plant simply for data storage); in increased vulnerability; in training; in maintenance; in expensive new secondary problems to solve. Information is a huge, increasing and mostly ignored manufacturing overhead, which needs to be planned, controlled and used as meanly as energy or any other expensive resource. It is in this light that the search for competitive advantage in manufacturing must be seen. It is *appropriate* technology which is the holy grail. That may imply more information technology than a company has at present but it may also mean less. The IAM/DTI/Kearney (1984) report found that the most effective users of IT were not necessarily the heaviest spenders. A notable characteristic of most world class companies is a healthy distrust of technology for its own sake.

> 'Lots of people think you need computers and automation for high-class manufacturing. You do need computers because of the sheer weight of information. But much more, you need simplification and integration, so that you can use the electronic tools. All the capital equipment in the world is useless without organization and people' (manager at IBM's Havanet plant).
>
> 'I believe a great deal of harm has been done by people pursuing the holy grail of CIM. It worries the hell out of me when people talk about "our goal being CIM". That's bullshit. My goal is making a more cost-effective product. If down the road that means more computer power, fine. If it means buying a second-hand press in California, I'll do that too' (manufacturing director, JCB).
>
> (Caulkin, 1988)

In the age of processing plenty, the pursuit of the zeros suggests that knowing what to leave out may be as important for computing advantage as what to leave in.

The framework

Implicit in the 'more is better' discussion of the latest technological tools (CIM, programmable automation, artificial intelligence) is the discouraging assumption that they are for big companies only. Large-scale automation, manufacturing automation protocol (MAP) and FMS are big-ticket items, well out of the investment and maintenance reach of smaller firms. Yet mass manufacturers, in particular consumer mass manufacturers, are only a tiny minority of producing firms, a fact which their high visibility disguises. Small and medium-sized job shops bulk largest in the overall manufacturing economy. They have just as much need and just as much to gain from better scheduling, better materials control, better design and better responsiveness, which together are the province of information strategy.

Most firms are unlikely to need, let alone to be able to cope with, thinking robots, automated storage and retrieval systems (AS/RS), FMS and automated guided vehicles (AGVs). Typically, working from the priority of reducing cycle time and increasing their responsiveness to customer requirements, they will require a measure of computing to improve scheduling, a simple signalling system to control materials, some CNC and/or a regrouping of existing machines to permit the processing of families of parts, a recasting of the clerical functions as internal services, design for manu-facture, some electronic links with customers and suppliers and a great deal of training. The tools are likely to be PCs and microprocessors rather than mainframes and massive centralized databases. The 'simplify-integrate-automate' approach to information technology applies to these 'frugal' methods and to the most ambitious CIM.

Indeed, the major initial gains from the successful introduction of IT ironically derive from the 'simplify–integrate' modules of the process rather than computerization itself. On the other hand, the simpler and more integrated a firm's processes, the easier and more attractive computerization subsequently becomes. Advanced IT-based technology can make good manufacturing performance better; it cannot on its own turn a bad factory into a good one. Current factory developments support this thesis, which explains, for instance, the differing patterns of automation in Japan and the USA. Many observers have been struck by the relatively slow growth until quite recently of large-scale automation in Japan, where despite extensive and encouraging experience with CNC, FMS and robots, fully computer integrated operations have been rarer than in the USA. Now that the major benefits of just-in-time production (simplification) and total quality (integration) have been won, however, the pace of automation appears to be increasing across a range of Japanese industries from cars to calculators. The aim, as the INSEAD research among others suggests, is not only to tighten the cost and quality screw; it is also to attack rivals on the new strategic battlegrounds of variety, flexibility and shorter response times.

Conversely, despite much initial enthusiasm there are few successful US manufacturers which ascribe their improved performance to better manufacturing hardware, and even fewer, apart from GM, which have publicly pinned their future hopes on it. IBM's revealing attitude to JIT has already been mentioned. Harley-Davidson, Tektronix and Hewlett Packard are also fans. Other companies have chosen quality as a suitable strategic driver to integrate corporation-wide improvement efforts, sometimes with striking effect. Ford's high roll, Xerox's fightback against formidable Japanese competition in copiers and Westinghouse's white collar productivity initiatives are exemplary cases of advanced companies using strategic quality goals to drive their technology policies. Excellent UK companies like JCB, Linn Products and the rejuvenated Lucas all place quality in its largest sense (that is, customer satisfaction) as their prime business concern.

Technological change

The crucial importance of such a broad framework is that it provides a perspective in which to situate technological change. Without this kind of perspective concepts like CIM are literally meaningless: mechanical abstractions devoid of intent or purpose. For this and other reasons, detailed technological prediction is not in itself a specially fruitful activity. It is useful, however, to assess some of the emerging technologies and automation trends within the larger strategic context, in which situation they tend to take on a rather different emphasis from that given in the technical literature.

Artificial intelligence

As with almost every other apparent technological 'breakthrough', artificial intelligence (AI) has promised more than it has actually delivered. So far it has been successful in specialized, well bounded areas such as medical diagnostics and the interpretation of seismological data. Typically, factory operations are much more diverse, have many more variables and enjoy less well defined boundaries. Consequently, although AU applications in machine diagnostics (such as maintenance and malfunction correction) are beginning to

appear, it will be a long time, if ever, before artificial intelligence can be widely applied to hybrid disciplines like process planning or scheduling. AI attractively embodies the theory of IT as technology transfer, distilling the best available knowledge in a form accessible by other people. In practice, the problem is not one of processing power: it lies in defining the problem to which the processing power is applied and in extracting the knowledge and skill of the expert in an explicit form which can be encoded within a computer system. Both these problems are pretty intractable in the undisciplined and loosely bounded world of manufacturing, severely limiting AI's strategic value, which at least for now is likely to be in less ambitious areas where competent humans are in short supply.

Office automation

The next frontier for productivity, cost containment and quality improvement is the office. Far too long neglected in the obsession with cutting the direct labour headcount, many offices are unreconstructed bastions of the old bad habits of excessive compartmentalization and tunnel vision, with predictable results in bureaucracy, delay and 'invisible inventory'. So far, offices have won no overall productivity gain from their spending on information technology, largely because, apart from finance departments and more recently drawing offices, office functions have scarcely begun to reflect the organization changes that it makes both possible and mandatory in order to yield full strategic effect.

This is especially true of Japan, where archaic and paper filled (although leanly manned) offices frequently present a notable contrast to the quiet hum of purposeful factories. This is not, however, a particularly comforting thought. It means that even at present high levels of overall efficiency many Far Eastern competitors still have large potential for IT improvement, and they know it. The signs are that having satisfactorily tamed the factory the Japanese are now turning their attention to the office. As George Dorman noted (Westinghouse case study), where in the past the steady stream of Japanese camera and notebook wielding plant visitors was eager to inspect factory processes and equipment, they now want to examine the latest office routines. One Japanese authority has confidently predicted that within ten years, Japan will have overtaken the West in office efficiency as it has done in the factory. The confidence is not surprising, and the levels to match are low. The strategic, business-led approach to automation which the Japanese have used in the factory is also ideally suited to office applications.

In time, information technology will suggest new ways of organizing offices, as it is beginning to do in production. Thomas Gunn (1987), for instance, suggests that computer integrated marketing will connect the business directly with its customers, enabling it to build up better profiles of users and their needs. Advanced manufacturing plants are already exploring ways of supplementing corporate marketing information by using their own customer links (after sales service, customer claims, assigning engineers to sales teams, analysing lost orders, and so on) to sharpen market response. Robb Wilmot (1988) foresees a 'centrifugal organization' in which the opportunities for innovation occur at the interface between customer and supplier at the periphery: networking becomes both a technology and a key organization process for channelling information to the high speed, high energy part of the company at the edge. This has obvious implications for R & D departments, themselves a

Case study: Westinghouse Electric Corporation 1989

(Excerpts from a speech by George C. Dorman, VP Human Resources, Westinghouse Electric Corporation)

'Every summer, we experience an invasion by the Japanese. They flock to American business firms, armed with tape recorders, cameras and a thousand questions. A large number of them visit Westinghouse each year.

'Until recently, our Japanese visitors were mainly interested in touring our factories – presumably to check on whether we were catching up. But now their focus is changing. These days, they want to know about productivity and quality procedures in our offices.

'This is a very significant change.

'It is clear that we are experiencing a fundamental shift in the economics of a productive enterprise. The ability to leverage the knowledge and expertise of our white collar activities has now become a dominant factor in our ability to compete in a worldwide marketplace.

'In Westinghouse, for example, information workers now represent two-thirds of our workforce – and over 75% of our total payroll costs. And those proportions are increasing. Our largest occupational group is engineers, followed by clerical people, salesmen and accountants.

'And guess what? These information workers are also a major key to quality performance. The studies of our Corporate Quality Centre show nearly two thirds of our costs of quality failures are due to white collar issues – not factory problems.

'So, in advanced nations, the burden of competition – both costs and quality competition – has shifted to the complex, high value added activities of our information workers. How do we manage this vitally important white collar improvement process?

continues

continued

'Our mission is to build value for our shareholders, our customers and our employees. The way we do it is by emphasizing total quality in everything we do. Eight years ago we founded the Westinghouse Corporate Quality Centre. Today, with 220 employees, it may be the largest activity of its kind in the world. The Centre works to install total quality culture and processes in every part of Westinghouse.

'Nearly two-thirds of the Centre's activities today are focused on high leverage white collar performance, in terms of both productivity and quality. We're still making impressive gains in factory performance – with JIT practices and so on. But the biggest total quality improvements are coming in the sales and order entry offices, the drafting rooms and the designer's cubicles... all the way through the business process to the treasury people who collect receivables.

'In one of our businesses, [a program to reduce cycle time] helped restructure and then automate order entry procedures... cutting the time needed to handle an electrical parts order from 28 hours to just ten minutes – and slashing cost per order by two-thirds at the same time.

'In a more complex example, the engineering design cycle for nuclear plant fuel reload assemblies was reduced from three years to 18 months – lowering costs by 25% – and allowing the engineering department to handle a 40% increase in workload with only 10% more people.

'For both instances the technique was the same: first streamline the process; then apply technology to implement the new, improved process. You can't do it in the reverse order – process must come first. In both cases, these were important white collar processes: one in marketing, the other in engineering.'

major and as yet unimproved factor in cutting new product lead time. For the moment, however, most firms have more than enough to go for in basic improvements in the areas of quality, reduction of cycle times, better R & D management and refocusing traditional departments to serve internal customers. These are all strategic targets without which the hard-won changes on the factory floor lose half their effect.

Supercomputers

The main issue in manufacturing is not raw computing power but defining and bounding complex problems such as plant scheduling and loading in order to bring processing power to bear. Until these problems are solved, ultra-fast, ultra-powerful and expensive supercomputers are unlikely to have a strategic impact on the factory. Currently, supercomputers are mainly used in immense number crunching applications like weather forecasting, academic scientific research, defence and computer generated special effects in films. They may also have potential in R&D (computer companies use them), although up to now their ability to output the figures in graphics form has been limited. Extraordinary jumps in processing power are now beginning to bring super-computing work stations into view, and some industry figures believe that the growth of shared systems, making increasing demands on central computer time, will provide an additional market. The prospect is seductive, but readers will recognize it as another solution of complexity. In particular software is expensive and hard to write, and there is currently little of it.

The peopleless factory/paperless office

Except in exceptional circumstances, neither is about to happen. The reasons are linked. The whole notion of the peopleless factory presupposes that there is an advantage in getting rid of the human presence. There is not. People can improvize, make intuitive jumps, learn from experience. In the end, people are the only source of competitive advantage: better informed, more experienced, more skilful people. As one observer has remarked, no one has yet seen a robot take part in a quality circle. It is certainly true that the growing use of IT changes the balance of the people needed in the factory, just as it switches the emphasis of the whole organization from product to service. The flexible, automated plant will need almost no blue collar workers in the traditional sense. It will, however, employ highly qualified people in an 'airline pilot' role: for most of the time carrying out routine checking and monitoring, but able to spring into action to override the machine where skilled judgement and improvization are called for.

For their part, the Japanese have concluded that at least for the next 20 years, factories will be manned by maintenance engineers and quality control people, if not by traditional direct operatives, running on increasingly refined JIT principles. They are therefore concentrating on automation as a 'man–machine system'. This typically bland formulation contains typically unbland implications. As one writer puts it:

> 'The traditional meaning of man–machine system put stress mainly on the interface with highly sophisticated machines (weapons). There,

he first priority was on machines that were complicated enough for a an to make a fatal mistake if the interfaces were not designed carelly. The new message on the man–machine system is in the point at, as automation becomes more and more common, the importance of the remaining people increases, and the plant of the near future will be planned and organized around the clearly defined functions of the remaining multipurpose workers (or blue collar engineers) first of all.'

(Abe, 1987)

It is clear enough by now that if factories will not be peopleless, nor will they be paperless. Parts of the operation positively demand computerization. For example, JIT converts find a main source of problems not in physical quality but in white collar error in the blizzard of paper accompanying small and often JIT deliveries. In such cases EDI positively aids both quality and productivity, and suppliers need little urging to operate. When Ford of Europe ran a pilot scheme for electronic invoicing, it was heavily oversubscribed. On the other hand, computers themselves create vast quantities of paper. So do laser printers, fax machines and photocopiers, all prime products of the IT revolution. (Not surprisingly, filing cabinets are another, more mundane growth industry.) In any case, as technology becomes more transparent and allows a better vision of the underlying goals, it is time to acknowledge that paper still has an important role in communications. For many applications it has a much lower total cost than IT. It is unthreatening; people feel comfortable with it and will willingly use it where they shy off screens and keyboards. This is especially (although not exclusively) true of the shop floor, where operator allergy to black boxes has contributed substantially to the failure of many MRP installations. For this and other reasons, such as pressures to simplify, computerized shop floor data entry systems look to have a limited future, and where they have been installed, the most advanced manufacturing companies (Volvo at Kalmar and Hewlett Packard almost everywhere) are taking them out rather than putting more in. Wilmot has coined the term 'papermation' to denote the hybrid systems which combine the best of the old technology with chosen elements of the new.

Downsizing

A final trend to consider is the increasing irrelevance of traditional computer size categories. Top end microprocessors are more than a match for low end minicomputers in performance; and perhaps more to the point, powerful microprocessors chained together can equal the power of mainframes at a tiny fraction of the cost. One respected industry observer, William F. Zachmann a market researcher at International Data Corporation, has proclaimed the era of the large computer at an end. He thinks that the twin movement towards the microprocessor ('the irresistible economies of small') and a restricted number of standard operating environments signals the end of the traditional mainframe on straight economic grounds, and the beginning of the second era of information processing based on modular, easily expandable processors and standard operating systems.

This obviously has organizational implications for manufacturing firms. Up to now, large mainframes and minis have made data processing a classically centralized function in the hands of technical staff. Single centralized databases and manufacturing systems built round them, such as CIM, also work in that sense. Many companies will want to keep things that way, not only because of their large existing investment in hardware, software and training, but also for control reasons. Central file servers and mooted diskless work stations, for instance, would make absolutely sure that the only information on the network was both authorized and access controlled.

Against that is the stifling effect of such a structure on creativity. Potentially the most powerful effect of IT is in liberating information and its use; not restricting it, but opening it up for application in new and unpredictable ways. The shape of the monolithic centralized corporation is so familiar that it is difficult to visualize an alternative. But huge advances in database processing, the concurrent development of large relational databases for different functions and the high-powered, relatively low cost hardware to run them, plus advances in controlling distributed databases, suggest other possibilities. The energizing of the marketing/R&D interface proposed by Wilmot is one example of how IT might be used to activate and bring to bear the information contained and unexploited in filing cabinets scattered round all organizations. There will be other ways to yoke structure to the thrust of technological advance, and those alert to them will benefit, if only by reducing the ever growing cost of the information overhead.

Strategy for the future

The world class movement

Some of these trends and technologies will turn out to be more significant than others. But companies will have to take a view on all of them, not only to decide which options to adopt themselves, but also to ponder the implications of their rivals doing so. The strategic importance of such technical decisions, together with their resource implications, would alone be enough to justify the presence on a company board of a director directly responsible for the information resource. Taken with the rest of the argument developed in this book – the centrality of information and the need for systems to be business not technology driven – the case becomes urgent and unanswerable. A word of warning, however: resist the temptation simply to change the name of the DP manager to chief information officer and leave it at that. This will be a new, cross-functional job, and the best candidate for it will not be a narrow DP specialist. Information management is too important and too broad in scope to be left to the DP department, which is largely why up till now the strategic promise and the actual achievement of information technology have remained so far apart. If companies wish to pay more than lip service to the importance of the information resource (and its cost), they must appoint as information boss someone with the vision to see through traditional corporate organization and competition structures as well as possessing the management clout to champion new solutions. He or she also needs to be able to weigh up improvements in processing through the use of IT against changes to the product, again using IT.

The formal recognition of the information officer's role is the more important since western manufacturing is poised on the cusp of historic change. Already some companies have pushed over the crest and are now driving hard down the learning curve towards world class competitiveness, based on shorter cycle time, greater responsiveness and total quality, with an information strategy to link and fine tune them. In 1987 Schonberger compiled a consciousness-raising roll of honour of '5–10–20' companies or plants or parts of plants which have achieved fivefold, tenfold or twentyfold reductions in manufacturing lead time. This all-American list comprised mainly high tech firms (although there was an honourable sprinkling of improvers in automobiles and components, chemicals, drapery fabric, tools, toiletries and cough drops). There are similar, though fewer, world class plants in Europe, not all of them foreign owned. But these firms, accustomed to rapid rates of change and exposed to the full rigours of global competition, are not typical. Similar achievements are needed just as much in the huge majority of companies in more mundane industries which have yet to come to terms with the new and infinitely more threatening manufacturing environment.

The word is beginning to get through. The accumulating success stories of manufacturing improvement are too consistent in pattern for the causes to be random. Better than exhortation, leading companies are spreading the gospel by belatedly demanding quality, better delivery performance and a measure of integration from their suppliers. Ford of Europe has an elaborate procedure for assessing vendors, which it divides into 'preferred', 'adequate' and 'inadequate' categories. The inadequate do not long remain on the list, although the sincerely repentant will get Ford's help to reform. Suppliers in their turn have no choice but to persuade their own vendors to march to the same drum.

But they are chasing a moving target. The Japanese suggest that the diffusion of JIT in its present form through suppliers to distributors is substantially complete. In Europe and the USA it has barely begun. The Japanese are now busy with the next developments: computerized just-in-time information systems to adapt simple repetitive factories and processes for more flexibility, and just-in-time product development by means of simultaneous engineering, assisted by techniques such as quality function development. In a world of manufacturing overcapacity the two developments go together: ever quicker innovation and segmentation are needed to maintain market share, while the near impossibility of depreciating each generation of plant and equipment within the shortening life-cycles of current products demands increasingly flexible manufacturing plants.

The role of IT has to be seen in this unforgiving perspective. IT for its own sake is pointless in the literal sense. No IT investment is worth making unless it contributes to the overall aim of making the business more competitive by increasing total value for the customer, whether by shortening response times, cutting defects in goods or services, or both. It is important to realize at the same time that bad IT investment can actually make all these matters worse.

The just-in-time business

The imperatives are simply stated. Since around 1980, manufacturing industry throughout the world has been faced with growing overcapacity (competition) and increasing demand for greater product variety. Hence the demand for flexibility and low cost; hence also the need to produce products just in time and in just the quantities for market. The model is the Japanese shoe manufacturer which has installed laser measuring devices in shoe shops to gauge the customer's foot size and transmit the information to the factory, which makes up the bespoke shoes overnight; or Toyota's vision of being able to build a new car to a customer's individual order and deliver it anywhere in Japan within six days (four in Tokyo). This is the just-in-time business: *business,* not just manufacturing, without the rocks in the river to disturb the flow, where the norm is no waste – no wasted time, no wasted quantity, since nothing is made without a customer order.

In this world of overcapacity, the real economies of scale are not in manufacturing as such (low costs are a given) but in satisfied customers. The costs of an extra new sale are enormous compared with the cost of repeat sales. Since anyone can copy features and (at least for a short time) price, this above all demands quality and reliability of product and service, which just happen to be the most difficult attributes for a competitor to duplicate. (Think of IBM: few people buy IBM for features or price. Or think of Ford: it has taken it eight years of preaching quality as job number one to lever quality levels to within range of Japan and West Germany.) The real barriers to industry entry are the 75-year investment in getting hundreds and thousands of people to live service, quality and customer problem solving at IBM, or the 150-year investment in quality at Procter & Gamble. These are the truly insuperable barriers to entry based on ironclad traditions of service, reliability and quality. This is the world for which manufacturing companies are installing IT systems today and for which manufacturing skill turns out to be an unexpected key to competitive advantage.

Case study: Boots

When Boots the Chemist's new store in Aberdeen, Scotland, opened for business on Monday 23 November 1992, it signalled the completion of a roll-out programme that created the largest EPOS (electronic point of sale) system outside of North America.

The project had taken six years and entailed expenditure of £70 million. The development of equipment and software to meet the need of Boots' 1100 stores was undertaken jointly with IBM UK whose 4680 store system had been at the heart of the installation. There are now 12,500 EPOS tills processing over 700 million customer transactions per annum.

The main benefits for Boots' customers are speed of service and itemized till receipts but there have also been major benefits to the company in the efficient and profitable management of the business.

Managing Director of Boots the Chemist, Gordon Hourston, had no doubts about the value of the investment in EPOS.

> 'The contribution to the company's net profitability has been very substantial. We have been able to improve so many aspects of the business – from operational improvements in the stores to more efficient automated stock control systems, detailed analysis of sales performance and the ability to target our product ranges and tailor our displays through better understanding of the customers' needs.'

Much of the focus until now has been on improving the bottom line by reducing costs and widening margins but there are still major benefits to come in terms of the feedback to the factories and to the sales generating activities.

Manufacturing as driver

For the flexible manufacturing processes at the centre of the firm are just one step in a just-in-time information and intelligence loop which links customer preference, superior product, suppliers, manufacturing, sales, distribution and retailer or dealer. At the start of the information age, just-in-time methods seemed simply the most rational means of managing materials and, more recently, information for manufacturing. But as their strategic implications are explored, manufacturing is more and more the model for the rest of the firm. Manufacturing, previously the neglected source of complexity and cost, becomes the shock force which drives quality, continuous improvement and timeliness through all the other functions.

Stretching it a bit? Consider Toyota. When it integrated its previously separate sales and production companies in the 1980s, the driving force was the discontent of manufacturing managers that performance in sales and distribution was not matching the ceaseless improvements in production. Manufacturing was taking a day to build a car; but it still took four weeks from customer order to delivery. In Toyota's subsequent reorganization, senior marketing people were quietly shifted sideways and their place taken by eager just-in-time manufacturing men who brought with them the urgent concern for product and process quality learned on the shop floor. The increase in Toyota's market share from 32 to 50% which followed was a tribute to the power of ideas, chiefly concentrated on cutting cycle time, developed through manu-facturing, not sales.

This then is the real context, and the challenge, for IT in manufacturing. The revolution is not, after all, the technology itself, but using it to measure up to ever more challenging business ambitions. In that sense, the concept of IT in manufacturing, like IT in finance or IT in marketing, is a distraction. There is merely better information management to keep the firm abreast of its world class aims. IT makes only the most routine and unimportant jobs easier. Developing a business plan, devising a manufacturing strategy, motivating people, designing a product that is more attractive than the rival firm's, building quality and experience: all these things call for the same rigorous analysis and imagination they always did. Computers can help, of course, but they provide no short cuts. In business, as Peter Drucker rightly says, 'Eventually everything degenerates into hard work.'

References

Abe K. (1986). How the Japanese see the future in JIT. In *JIT Manufacturing* (Voss C. A. ed.). Proceedings of the First International Conference on Just-in-Time Manufacturing. IFS publications, Springer Verlag

Agurén S., Bredbacka C., Hansson R., Ihregren K. and Karlsson K. G. (1984). *Volvo Kalmar Revisited: Ten Years of Experience.* Efficiency and Participation Development Council SAF LO PTK

Caulkin S. (1988). Britain's best factories. *Management Today*, September

DTI/Logica/Price Waterhouse (1988). *Financial Times*, May 6

Gunn T. G. (1987). *Manufacturing for Competitive Advantage.* Ballinger Publishing Co.

Kearney A. T. (1984). *The Barriers and Opportunities of IT – a Management Perspective.* Institute of Administrative Management and the DTI

Lyytinen K. and Hirschheim R. (1987). Information systems failures; a survey and classification of the empirical literature. In *Oxford Surveys in Information Technology*, **4** (Zorckoczy P. I. ed.). Oxford: Oxford University Press

McCormick and Paget (1987). *Positioning Corporate Staff for the 1990s.* American Productivity Center/Cresp

Schonberger R. (1986). *World Class Manufacturing.* Free Press

The software trap – automate or else. *Business Week.* May, 1988

Wilmot R. W. (1988). Seizing the initiative: the strategic use of IT. *IT Perspectives Conference: The Future of Information Technology.* Computer Weekly Publications

Glossary

ACARD	Advisory Council on Applied Research and Development (UK)
AGV	Automatic Guided Vehicle
AI	Artificial intelligence
AMT	Advanced manufacturing technology
AS/RS	Automatic storage and retrieval systems
AV	Audio visual
BOM	Bill of materials
CAD	Computer aided design
CAE	Computer aided engineering
CAI	Computer aided inspection
CAM	Computer aided manufacturing
CAPM	Computer aided production management
CAPP	Computer aided production planning
CASE	Computer aided software engineering
CAT	Computer assembly and test
CBT	Computer based training
CFM	Continuous flow manufacturing
CIB	Computer integrated business
CIM	Computer integrated manufacturing
CM	Condition monitoring
CNC	Computer numerically controlled
CRLC	Customer resource life critical
DNC	Direct numerical control
DP	Data processing
DRP	Distribution resource planning
DTI	Department of Trade and Industry
ECAM	Electronic CAM program (US)
EDI	Electronic data interchange

EETPU	Electrical, Electronic, Telecoms and Plumbing Union
EI	Employee involvement
EOQ	Economic order quantities
EPOS	Electronic point of sale
ESPRIT	European strategic programme for R&D in IT
FA	Flexible automation; factory automation
FAST	Federation Against Software Theft
FM	Facilities management
FMS	Flexible manufacturing systems
GT	Group technology
HCIM	Human and computer integrated manufacturing
ICAM	Integrated CAM program (US)
IGES	Initial graphics exchange specification
ISO	International Standards Organization
IT	Information technology
JIC	Just-in-case
JIT	Just-in-time
Kaizen	Continuous improvement
Kanban	Simple record cards
LAN	Local area network
MAP	Manufacturing automation protocol
MITI	Ministry of International Trade and Development (Japan)
MPS	Master production schedule
MRP/II	Material requirements planning/manufacturing resource planning
NC	Numerically controlled
NCC	National Computing Centre
OPT	Optimized production technology
OSI	Open standards interconnection
OTED	One-touch exchange of die
PA	Programmable automation
PC	Personal computer
PI	Process information
PLC	Programmable logic controller
PPM	Planned preventative maintenance
QC	Quality control
QFD	Quality function development
ROI	Return on investment
SAA	Systems application architecture (IBM)
SET	Standard d'échange et transfert (Airbus)
SMED	Single-minute exchange of die
SPA	Software Publishers Association
SPC	Statistical process control
SQC	Sight quality control
TOP	Technical office protocol
TPM	Total preventive maintenance
TQC	Total quality control
TQM	Total quality management
UNIX	Computer operating system developed by AT&T
WAN	Wide area network
WCDP	Work cell device programming (of CNC)
WIP	Work in progress

Bibliography and further reading

Abe K. (1987). How the Japanese see the future in JIT. *JIT Manufacturing*. Berlin: IFS, Springer-Verlag

ACARD (1983). *New Opportunities in Manufacturing: the Management of Technology*. London: HMSO

Ackerman L. (1984). The flow state: a new view of organizations and managing. In *Transferring Work* (Adams J., ed.). Miles River Press

Adam E. E. and Ebert R. J. (1989). *Production and Operations Management*. London: Prentice Hall

Advanced Manufacturing Systems Group (1985). *Advanced Manufacturing Technology; The Impact of New Technology on Engineering Batch Production*. NEDO

Agurén S. *et al.* (1984). *Volvo Kalmar Revisited: Ten Years of Experience*. Efficiency and Participation Development Council SAF LO PTK

Akao J. (ed.) *Quality Function Deployment*. Cambridge, Mass: Productivity Press

Andreasen M. M. and Hein. *Integrated Product Development*. Bedford: IFS

Angel, I. and Smithson, S. (1989). *Managing Information Technology: A Crisis of Confidence*. LSE Department of Information Technology Working Paper number 20. London: London School of Economics

Bendall A., Disney J. and Priomore W. A. (eds.). *Taguchi Methods: Applications in a World Industry*. IFS Publications, Springer-Verlag

Bowman D. J. , and Bowman A. C. *Understanding CAD/CAM*. Indianapolis: Sams

Browne J., Harhen J. and Shivnan J. (1988). *Production Management Systems: a CIM Perspective*. Wokingham: Addison-Wesley

Buchanan D. and Boddy D. (1983). *Organisations in the Computer Age*. Aldershot: Gower

Buchanan D. (1986). Management objectives in technical change. In *Managing the Labour Process* (Knights D. and Willmott H., eds). Aldershot: Gower

Buffa E. S. and Sarin R. K. (1987). *Modern Production/Operations Management*. New York: Wiley

Business Week (1988). The software trap: automate – or else. May 9

Butler Cox Foundation (1990). *Getting value from information technology*. Research Report 75. June. London: Butler Cox

Caulkin S. (1986). Why Daewoo works harder. *Management Today*, July

Caulkin S. (1987). ICL's Lazarus act. *Management Today*, January

Caulkin S. (1988). Britain's best factories. *Management Today*, September

Chase R. B. and Aquilano N. J. (1989). *Production and Operations Management*. Irwin

Coleman T. and Jamieson M. (1991). *Information Systems: Evaluating Intangible Benefits at the Feasibility Stage of Project Appraisal*. Unpublished MBA thesis. London: City University Business School

Coopers & Lybrand (1988). *The Security of Network Systems: A Report on Behalf of the Commission of the EC*

Coopers & Lybrand/BIM (1988). *Managers and IT Competence*

Cullen J. and Hollingum J. *Implementing Total Quality*. IFS Publications

De Meyer, A. *et al.* (1987). *Flexibility: the next competitive battle*. INSEAD Working Paper 86/31

Dempsey, P. (1986). Breaking new ground in JIT. *JIT Manufacturing*. Berlin: IFS, Springer Verlag

Dillworth J. B. (1988). P*roduction and Operations Management*. Random House

Drucker, P. (1987). *The Frontiers of Management*. Heinemann

Drucker, P. (1988). The coming of the new organization. *Harvard Business Review*, January–February

DTI (1987). *Towards Integration*

Garvin D. (1988). *Managing Quality; The Strategy and Competitive Edge*. Free Press

Gilmour D. (1987). JIT in action. *QA News*, June

Goldratt E. and Cox J. (1984). *The Goal: Excellence in Manufacturing*. North River Press

Greenwood N. R. (1988). *Implementing Flexible Manufacturing Systems*. MacMillan

Griffiths P. (ed.) (1987). *The Role of Information Management in Competitive Success*. Pergamon Infotech, April

Grindley K. (1991). *Managing IT at Board Level: the hidden agenda exposed*. London: Price Waterhouse/Pitman

Groover M. P. (1987). *Automation, Production Systems and Computer Integrated Manufacturing*. Prentice Hall

Gunn T. G. (1982). The mechanization of design and manufacture. *Scientific American*, September

Gunn T. G. (1987). *Manufacturing for Competitive Advantage*. Ballinger Publishing Co.

Harrison M. (1990). *Advanced Manufacturing Technology Management*. London: Pitman

Hartland-Swann J. (1986). Three steps to CIM. *Industrial Computing*, April

Hartland-Swann J. (1987). How much integration? *Industrial Computing*, February

Hawken P. (1983). *The Next Economy*. Bantam

Hay E. J. (1988). *The JIT Breakthrough: Implementing the New Manufacturing Basics*. John Wiley and Sons

Hayes R. H. and Wheelwright S. C. (1984). *Restoring our Competitive Edge: Competing Through Manufacturing*. John Wiley & Sons

Hayes R. H., Wheelwright S. C. and Clark K. (1988). *Dynamic Manufacturing: Creating the Learning Organization*. Free Press

Hill T. (1985). *Manufacturing Strategy: the strategic management of the manufacturing function*. Macmillan

Hochstrasser B. and Griffiths C. (1991). *Controlling IT Investments: Strategy and Management*. London: Chapman and Hall

Hollingum J. (1987) *Implementing an Information Strategy in Manufacture*. IFS, Springer Verlag

IBM (1987). *Computer Integrated Manufacturing: the IBM Experience*

IMechE (1987). *Information Pack on CIM*. Information Pack 4

Ingersoll Engineers (1985). *Integrated Manufacture*. IFS, Springer Verlag

Jaikumar R. (1986). Post-industrial manufacturing. *Harvard Business Review*, November–December

Johne F. A. and Snelson P. A. (1987). *Product Policy and Development in Manufacturing Firms*. London: City University Business School

Johne F. A. and Snelson P. A. (1987). *Success Factors in New Product Development*. London: City University Business School

Johne F. A. and Snelson P. A. (1987). *Product Development Practices in Large US and UK Firms*. London: City University Business School

Kaplan R. S. (1987). Yesterday's accounting undermines production. *Harvard Business Review*, July–August

Kearney A. T. (1984). *The Barriers and Opportunities of Information Technology – a Management Perspective*. Orpington: Institute of Administrative Management

Kearney A. T. (1987). *Corporate Organisation and Overhead Effectiveness Survey*. Orpington: Institute of Administrative Management

Keen P. G. W. and Gooding G. (1987). *Strategic Investments in Information Technologies; Leading Versus Following; An Example from Videoconferencing*. International Center for Information Technologies

Kelman A. (1985). *Computer Fraud in Small Businesses*. London: Economist Intelligence Unit

Kimmerly W. (1986). Managing the risks of installing CIM. *Computerworld*, October 13

Kobayashi, I. *20 Key Ways to Workplace Improvement*. Productivity Press

Kobler Unit (1987). *Does Information Technology Slow You Down?* London: Kobler Unit for the Management of Information Technology, November

Kobler Unit (1990). *Regaining Control over IT Investments*. London: Kobler Unit

Laudon K. C. and Turner J. (ed.) (1989). *Information Technology and Management Strategy*. Prentice Hall

Lyytinen K. and Hirschheim R. (1987). Information systems failures; a survey and classification of the empirical literature. In *Oxford Surveys in IT* Vol 4, 1987, Zorckoczy (ed.). Oxford: Oxford University Press

McCormick and Paget. (1987). *Positioning Corporate Staff for the 1990s*. American Productivity Center/Cresp

New C. C. and Myers A. (1987). *Managing Manufacturing Operations in the UK 1975–1985*. BIM

Owen T. (1987). *Robots out of Wonderland: How to Use Robots in the Age of CIM*. Cranfield Press

PA Consulting Group (1990). *The Impact of the Current Climate on IT – The Survey Report*. London: PA Consulting Group

Parker M. M., Benson R. J. and Trainor H. E. (1988). *Information Economics*. London: Prentice Hall

Porter M. and Millar V. (1991). How information gives you competitive advantage. In *Revolution in Real Time: Managing Information Technology in the 1990s* (McGowan W.) (preface). Boston: Harvard Business School Press: 59–82

RIPA/Hoskyns (1989). *Investing in Quality: Managing the Technological Change of the 90s*. London: RIPA/Hoskyns Joint Symposium

Schonberger R. J. (1967). Frugal manufacturing. *Harvard Business Review*, September–October

Schonberger R. J. (1982). *Japanese Manufacturing Techniques*. Free Press

Schonberger R. J. (1987). *World Class Manufacturing Casebook: Implementing VIT and TOC*. Free Press

Schonberger R. J. (1987). *World Class Manufacturing*. Free Press

Schweitzer J. A. (1986). *Computer Crime and Business Information: A Practical Guide for Managers*. Elsevier

Scott M. (1991). *The Corporation of the 1990s*. Oxford: Oxford University Press

Sheridan T. (1986). How to account for manufacturing. *Management Today*, August

Shingo, S. (1985). A *Revolution in Manufacturing: the SMED System*. Cambridge, Mass.: Productivity Press

Shingo, S. *A Study of the Toyota Production System*. Cambridge, Mass: Productivity Press

Skinner W. (1985). *Manufacturing: the Formidable Competitive Weapon*. John Wiley & Sons

Skinner W. (1988). What matters to manufacturing. *Harvard Business Review*, January–February

SRI (1976). *Industrial Automation in Discrete Manufacture*

Strassman P (1990). *The Business Value of Computers*. New Canaan: The Information Economics Press

Twiss B. and Goodridge M. (1989). *Managing Technology for Competitive Advantage*. London: Pitman

US Congress (1984). *Computerized Manufacturing Automation*. Office of Technology Assessment

Valery N. (1987). The factory of the future. *The Economist,* May 30

Voss C. A. (ed.) (1987). *Just-in-Time Manufacture*. IFS

Wall Street Journal (1988). Factory of the future becomes a vision of the past. 1 September

Warner T. N. (1987). Information technology as a competitive burden. *Sloan Management Review*, Fall

Waters C. D. J. (1991). *An Introduction to Operations Management*. Wokingham: Addison-Wesley

Wickens P. (1987). *The Road to Nissan*. Macmillan

Wild R. (1990). *Essentials of Production and Operations Management*. Cassell

Willcocks L. (1990). Theme issue: the evaluation of information systems investment – an introduction. *Journal of Information Technology*, **5**(4)

Willcocks L. (1992). Evaluating information technology investment: research findings and reappraisal. *Journal of Information Systems,* **2**(2)

Willcocks L. and Lester S. (1991). Information system investments: evaluation at the feasibility stage of projects. *Technovation*, **II**(5) 283–302

Willcocks L. and Lester S. (1992). *Of Capital Importance: Evaluation of IS Investments*. London: Chapman & Hall

Willcocks L. and Mason D. (1992). *Computerising Work: People, Systems Design and Workplace Relations* (2nd edn). Oxford: Blackwell Scientific

Willman P. (1987). Industrial relations issues in advanced manufacturing technology. In *The Human Side of Manufacturing Technology* (Wall T., Clegg C. and Kemp N.) Chichester: John Wiley & Sons

Wilmot, R. W. (1988). Seizing the initiative: the strategic use of IT. *IT Perspectives Conference: The Future of Information Technology*. Computer Weekly Publications

Wilson T. (1991). Overcoming the barriers to the implementation of information systems strategies. *Journal of Information Technology*, **6**(1), 39–44

Wilson G., Millar M. G., and Bendell T. *Taguchi Methodology with Total Quality*. IFS Publications

Zuboff S. (1988). *In the Age of the Smart Machine*. Oxford: Heinemann

Index